T0350967

A Contractor's Guide to Planning, Scheduling, and Control

WILEY

A Contractor's Guide to Planning, Scheduling, and Control

Len Holm

WILEY

This book is printed on acid-free paper.

Copyright © 2022 by John Wiley & Sons, Inc. All rights reserved

Published by John Wiley & Sons, Inc., Hoboken, New Jersey

Published simultaneously in Canada

No part of this publication may be reproduced, stored in a retrieval system, or transmitted in any form or by any means, electronic, mechanical, photocopying, recording, scanning, or otherwise, except as permitted under Section 107 or 108 of the 1976 United States Copyright Act, without either the prior written permission of the Publisher, or authorization through payment of the appropriate per-copy fee to the Copyright Clearance Center, 222 Rosewood Drive, Danvers, MA 01923, (978) 750-8400, fax (978) 646-8600, or on the web at www.copyright.com. Requests to the Publisher for permission should be addressed to the Permissions Department, John Wiley & Sons, Inc., 111 River Street, Hoboken, NJ 07030, (201) 748-6011, fax (201) 748-6008, or online at www.wiley.com/go/permissions.

Limit of Liability/Disclaimer of Warranty: While the publisher and author have used their best efforts in preparing this book, they make no representations or warranties with the respect to the accuracy or completeness of the contents of this book and specifically disclaim any implied warranties of merchantability or fitness for a particular purpose. No warranty may be created or extended by sales representatives or written sales materials. The advice and strategies contained herein may not be suitable for your situation. You should consult with a professional where appropriate. Further, readers should be aware that websites listed in this work may have changed or disappeared between when this work was written and when it is read. Neither the publisher nor authors shall be liable for any loss of profit or any other commercial damages, including but not limited to special, incidental, consequential, or other damages.

For general information about our other products and services, please contact our Customer Care Department within the United States at (800) 762-2974, outside the United States at (317) 572-3993 or fax (317) 572-4002.

Wiley publishes in a variety of print and electronic formats and by print-on-demand. Some material included with standard print versions of this book may not be included in e-books or in print-on-demand. If this book refers to media such as a CD or DVD that is not included in the version you purchased, you may download this material at http://booksupport.wiley.com. For more information about Wiley products, visit www.wiley.com.

Library of Congress Cataloging-in-Publication Data is Available:

ISBN 9781119813521(hardback)
ISBN 9781119813545(ePDF)
ISBN 9781119813538 (ePub)

Cover Design: Wiley
Cover Image: © kolderal/Getty Images

SKY10032113_122221

Contents

PART I   Introductory Topics   1

Chapter 1   Introduction   3

Chapter 2   Construction Management   15

PART II

Planning   33

Chapter 3

Preconstruction   35

Chapter 4

Schedule Planning   53

Chapter 5

Lean Construction Planning 79

Chapter 6

Contract and Time Considerations 95

List of Companion Website Materials

The following documents are available on the companion website for *A Contractor's Guide to Planning, Scheduling, and Control*, (Wiley, 2021). Reference www.wiley.com/go/holm/PlanningSchedulingAndControl

This first list is made available to students and instructors:

- Case study site logistics plan
- Sample preconstruction agreement
- Live schedule templates, including Excel three-week look-ahead schedule, submittal schedule, and expediting schedule – three files
- Live Excel estimating forms utilized in *Construction Cost Estimating*
- Live case study MS Project summary schedule, Figure 7.5
- Construction trade responsibilities
- To-do list
- Calendar schedule
- Case study progress photographs – three files 60 photographs

The following are also made available to instructors:

- *Instructor's Manual*, complete with answers to all of the review questions and many of the exercises, including spreadsheet answers to many of the exercises.

- A select group of case studies from *101 Case Studies in Construction Management* are included in the book and potential solutions, or *Suspects*, are also included in the *Instructor's Manual*.
- Power Point lecture slides for all 18 chapters – 18 separate files, almost 550 slides in total
- Select case study drawings
- Select case study specifications
- Case study detailed construction estimate, including jobsite general conditions and subcontractor list
- Case study detailed construction schedule
- Sample specialty contractor schedule for the case study project
- Sample residential project schedule
- Sample heavy civil project schedule
- Alternate flowchart work breakdown structure
- Detailed case study cost-loaded schedule
- Case study pay request schedule of values
- Sample subcontract agreement for the case study project

List of Figures and Tables

FIGURES

TABLES

Preface

Some authors, college professors, and even construction professionals think of the terms, processes, and documents used for 'planning' and 'scheduling' synonymously – as if they are the same thing – which they are not. In this book they each receive their own separate section, along with project controls, which is implementation of the plan and schedule. Each of the sections also include several supporting chapters.

A built environment professional learns to be a scheduler by scheduling, similar to estimating. You cannot completely get it from reading a book or attending a class. The book provides a good background and introductory topics, but the best way to teach it, and learn it, is through the use of examples and exercises. There are many of each of these in each chapter in the book. I have included almost 40 boxed-in examples of schedule situations spread throughout the book. Some of these are good examples of construction scheduling and some of them are not-so good. Often we learn more from our mistakes than from our successes. All of these are taken from projects that I have participated in professionally. Some of these examples will also be expanded on as well with end-of-chapter applied exercises.

With this book I am intentionally staying away from excessive academic jargon and not getting into the weeds with respect to scheduling terminology. This is not a book that was developed as someone's doctoral dissertation; it is rather a tool for students and professionals alike to learn more about construction scheduling. The focus of this book is on the construction contractor, not the scheduling consultant or claims consultant. Much of the academic scheduling terminology is not used out on the jobsite, but these concepts have been included and defined where appropriate and several good scheduling books have been included in the book's reference appendix if the reader wishes to pursue these concepts further. This book is geared for construction management and construction engineering scheduling students and construction professionals wanting to enhance their scheduling capabilities. The book's focus is on practical construction scheduling tools – those that help contractors build the project.

Most of the scheduling related textbooks use examples with activities A, B, and C, whereas this book will use actual construction activities, such as form, rebar, and place concrete, or wall framing; rough-in mechanical, electrical, and plumbing; test and inspect; and insulation and drywall so that the student learns the schedule logic along with means and methods of construction, as well as the mechanics and calculations of scheduling. The narrative and examples also use real craft workers, such as carpenters and electricians, and real subcontractors, such as earthwork and drywall, etc. After finishing the book the reader will have learned many important additional construction management topics, beyond planning, scheduling, and controls.

ACKNOWLEDGMENTS

To supplement typical academic coverage of construction scheduling, this book includes a practical construction perspective stemming not only from my almost 50 years of construction experience, but also from input offered by scores of construction professionals and friends. These practitioners have reviewed chapter drafts and commented on countless figures, tables, and exercises. Their experience is very much appreciated, for without them this would just be another college textbook. It would be difficult to list all of the people I need to thank, but I especially want to recognize:

- Eddie Baker, Construction Executive, Hermanson Company, LLP, Mechanical Contractor
- Reid Bullock, Senior Estimator and Preconstruction Manager, Compass Construction Company, General Contractor
- Sarah Elley, Project Architect, TCA Architecture, Planning, and Design
- David Holm, Project Engineer, Pence Construction, General Contractor
- Bob Ironmonger, Construction Manager and Scheduler, retired, lifelong friend, classmate, and coworker
- Cody Klansnic, Project Engineer, Lease Crutcher Lewis, General Contractor, for his research on the case study project
- Christian LaRocco, Real Estate Developer, MJR Development
- Amir Mahmoudi, Senior Project Scheduler, Clark Construction Group, LLC
- The City of Pasco, Washington, especially the Fire Department and Chief Bob Gear
- David Robison, Principal, Strategic Construction Management, Inc.
- Chengyi Zhang, Assistant Professor, Department of Civil and Architectural Engineering, University of Wyoming

I would also like to thank Kapil Devan, Kirk Hochstatter, and Lucky Pratama, recent graduate and doctoral students at the University of Washington, for their research and contributions to the book, select figures, and the instructor's manual.

There is a complete instructor's manual available on the companion website with answers to all of the review questions and many of the exercises. An appendix to the book also includes five case studies borrowed from *101 Case Studies in Construction Management* that pertain to planning, scheduling, and control. This is an excellent economical book that complements many construction management topics. Potential solutions, or *Suspects*, for these five case studies are also included with the instructor's manual.

If you have any questions about the material, or recommendations for changes for future editions, please feel free to contact the publisher, Wiley, or me direct at holmcon@aol.com. I hope you enjoy my connection of an academic study of construction scheduling to what we practice at the construction jobsite.

—Len Holm

List of Abbreviations

2D, 3D, 4D, 5D	two-dimensional (length and width), three-dimensional (depth), four-dimensional (time), and five-dimensional (cost)
AB	anchor bolt
ABC	activity-based costing
ABR	activity-based resourcing
ACE	assumptions, clarifications, and exclusions (bid and/or contract document)
ACP	asphalt
ACT	acoustical ceiling tile, formerly asbestos ceiling tile
ACWP	actual cost of work performed (earned value)
ADA	Americans with Disabilities Act (law or building codes)
ADM	arrow diagramming method
ADR	alternative dispute resolution
AEC	architecture, engineering, and construction (design-build company type)
AFF	above finish floor (dimension or height)
AGC	Associated General Contractors of America
AHJ	authority having jurisdiction, often city building department
AHU	Air-handling unit (mechanical equipment)
AIA	American Institute of Architects
AKA	also known as
Allow	allowance
Alt	alternate or alternative
AOA	activity on arrow scheduling method

AON	activity on node scheduling method
AQWP	actual quantity of work performed (earned value)
Arch	architect or architectural
ASI	architect's supplemental instruction, AIA form G710
Asst	assistant or assist
BC	back charge
BCAC	budgeted cost at completion (earned value)
BCF	bank cubic feet
BCWP	budgeted cost of work performed (earned value)
BCWS	budgeted cost of work scheduled (earned value)
BCY	bank cubic yard
BE	built environment
BF	board foot (lumber); also backfill
BL	baseline
BIM	building information models or modeling
BOT	build – operate – transfer (delivery method), or bottom (also BTM)
BQAC	budgeted quantity at completion (earned value)
BQWP	budgeted quantity of work performed (earned value)
BQWS	budgeted quantity of work scheduled (earned value)
BTR	better (lumber grading, for example *DF#2* and BTR)
BTU	British thermal units (heat)
BY	bank yard (cubic yard of dirt before excavation)
CA	construction administration (architect's role during construction), also carpenter
CAD	computer-aided design
CD(s)	construction document(s) (design phase or drawings)
CE	civil engineer or engineering, also construction engineering or chief estimator
CEO	chief executive officer
Cert	certificate
CF	cubic foot or cubic feet
CFO	chief financial officer
CIP	cast-in-place (concrete)
CM	construction manager or management
CM	cement mason, concrete finishing craftsman
CM/GC	construction manager/general contractor (also known as *GC/CM* or *CMAR*)
CMA	construction manager agency (delivery method)
CMAR	construction manager at-risk (delivery method); also *CM/GC*
CO	change order, county, or cased opening
C-O	close-out
C of O	certificate of occupancy

Contractor	general contractor or subcontractor
COO	chief operating officer, also construction operations officer
COP	change order proposal
CP	cost proposal, cost plus, carpet (also *CPT*), or critical path
CPFF	cost-plus fixed fee
CPI	cost performance index (earned value), also consumer price index
CPM	critical path method, also critical path network (CPN)
CPPF	cost-plus percentage fee (similar to *T&M*)
CPT	complete or carpet (also *CP*)
Coor(n)	coordination (process or drawings)
CSI	Construction Specifications Institute
CV	cost variance (earned value)
CY	cubic yard
D	dimension, diameter (also *dia*), dryer, or depth
DB or D-B or D/B	design-build (delivery method)
DBB	design-bid-build (delivery method)
DBO	design build operate
DBOM	design build operate maintain
DD	design development (documents or phase)
Demo	demolition
Demob	demobilization
DFH(W)	doors, frames, and hardware
DL	direct labor (estimate or cost area)
DM	direct material (estimate or cost area)
Doc(s)	document(s)
DRB	Dispute Resolution Board
E	engineer, east, exit, existing, or electrical
EA	each
ECD	estimated completion date (schedule)
EE	electrical engineer
EF	early finish (schedule); also exhaust fan
EMR	experience modification rating (safety)
EPC	engineering, procurement and construction (design build delivery)
ES	early start (schedule)
Est	estimate
ETA	estimated time of arrival
EV	earned value
EVM	earned value method
EW	each way or East-West
F	furniture, Fahrenheit, or fire protection

FCV	forecasted cost variance (earned value)
FE	field engineer or fire extinguisher
FF	finished floor (elevation); also finish to finish (schedule); also free float (schedule)
FF&E	fixtures, furniture, and equipment (often owner supplied)
FIC	furnished and installed by contractor
FIO	furnished and installed by owner
FOB	freight on board or free on board (material delivery location)
FOIC	furnished by owner and installed by contractor, also *OFCI*
FS	finish to start, also fire station
FSV	forecasted schedule variance (earned value)
FV	future value (time value of money)
FT	foot or feet
GC	general conditions, also general contractor
GC/CM	general contractor/construction manager delivery method, see also *CM/GC* and *CMAR*
Gen or general	general contractor
Geo	geotechnical (report or engineer); also known as soils engineer or report
GF	general foreman (similar to an assistant superintendent)
GMP	guaranteed maximum price, estimate or contract
GWB	gypsum wall board (also sheetrock or wall board or drywall)
H	height (also *HT*) or horizontal
HM	hollow metal (door or frame)
HO	home office
HOOH	home office overhead
HR	human resources, also hour
HSS	hollow structural section, formerly tube steel (*TS*)
HT	height (also *H*)
HVAC	heating, ventilating, and air conditioning (mechanical system or contractor)
IJ	ij nodes (activity on arrow nomenclature)
IPD	integrated project delivery
ITB	instructions to bidders, also invitation to bid
IW	ironworker (craftsman)
JIT	just-in-time (material deliveries)
JV	joint venture
LDs	liquidated damages
LEED	Leadership in Energy and Environmental Design (sustainability)
LF	late finish (schedule); also lineal or linear feet
LLC	limited liability company or corporation
LLP	limited liability partnership

LOI	letter of intent
LR	lien release
LS	lump sum (cost estimate, bid, contract, agreement, or process); also life safety (drawing); also late start (schedule)
LSM	linear scheduling method
M&E	mechanical and electrical (contractors or designers)
M&M	means and methods
MACC	maximum allowable construction cost, similar to *GMP*
MBF	thousand board feet (dimensional lumber measure)
MCC	Mountain Construction Company (fictitious case study GC), also motor control center (electrical equipment)
MEP	mechanical, electrical, and plumbing (systems or contractors)
MH(s)	man-hour(s) or man hole (sanitary sewer)
Mo(s)	month(s)
Mob	mobilization
MS	Microsoft
MUP	master use permit (entitlement permit)
NA or N/A	not applicable or not available
NCR	nonconformance report
NIC	not-in-contract, also not included
No	number (also #), or north
NTP	notice to proceed
O&M(s)	operation and maintenance manual(s)
OAC	owner-architect-contractor (commercial construction project meeting)
OE	operating engineer, also owner's equity
OFCI	owner-furnished contractor-installed, also *FOIC*
OH	overhead
OH&P	overhead and profit (also known as fee)
OIC	officer-in-charge
OM	order of magnitude (cost estimate)
OSHA	Occupational Safety and Health Administration
OT	overtime
P3 or P6	Primavera Project Planner (scheduling software system)
PDM	precedence diagram method (schedule)
PE	project engineer, pay estimate, professional engineer, or project executive
PERT	program evaluation and review technique (scheduling method)
PEx	project executive, also *PE* or *PX*
PL	punch list (also *punch*), plate, plastic laminate (also *plam*)
PM	project manager or management
PO	purchase order or project owner
PPE	personal protective equipment

PPP or P3	public-private partnership (delivery method) or Primavera Project Planner
PPP	pollution protection plan
PR	payment request, pair (doors); or public relations
Precon	preconstruction (services or contract or fee)
Prefab	prefabricated
Punch	punch list, also *PL*
PV	present value (time value of money)
PX	project executive, also *PE* or *PEx*
QA	quality assurance
QC	quality control
QE	quality engineering
QTO	quantity take-off
QTY	quantity, also **Q**
Rebar	concrete reinforcement steel
Recap	cost recapitulation sheet (estimating)
RFI	request for information, or request for interpretation
RFP	request for proposal
RFQ	request for qualifications, also request for quotation
R/I	rough-in
ROM	rough-order-of-magnitude (cost estimate)
ROT	rule of thumb
S	structural, south, supply, or survey (drawing)
Schd	schedule
SD(s)	schematic design (documents or phase); also smoke detector, soap dispenser, or storm drain
SF	square foot or square feet, also start to finish schedule relationship
SFCA	square foot of contact area
SFF	square foot of floor
SFW	square foot of wall
SHT(s)	sheet(s) (plywood)
Sim	similar
SIPS	short interval schedules; structurally insulated panels; or street improvement permit
SOG	slab-on-grade (concrete)
SOMD	slab on metal deck (concrete composite slab)
SOV	schedule of values (estimate or pay request)
Spec or specs	specifications, also speculation
SPI	schedule performance index (earned value)
SPM	senior project manager
Sprinks	fire sprinklers
SQ	square (100 square feet, roofing measure)

SS	start to start (schedule), also stainless steel or sanitary sewer
STP	Superintendent Training Program (part of *AGC*)
S/U	start-up
Sub(s)	subcontractor(s)
Subm	submittal
Super or Supt	superintendent
Supers	*Construction Superintendents, Essential Skills for the Next Generation* (textbook)
SV	schedule variance (earned value)
SWPPP	stormwater pollution protection plan
SY	square yard
T	thermostat, time, thickness, title sheet (drawing), topography (drawing), or ton
T&B	top and bottom
T&M	time and materials (contract or billing); similar to *CPPF*
TBD	to be determined
TC	tower crane
TCA	case study architecture firm
TCO	temporary certificate of occupancy
TESC	temporary erosion and stormwater control
TF	total float (schedule)
TI	tenant improvement
TJI	Truss Joist International (engineered lumber/I beam)
TN or Ton	tonnage (**2,000** pounds)
TNG	tongue and groove (also T&G)
TVD	target value design
TVM	time value of money
Typ	typical
UMH	unit man-hours
UNO	unless noted otherwise, also unless otherwise noted (UON)
UP	unit price
US or USA	United States of America
USACE	United States Army Corp of Engineers
USGBC	United States Green Building Council (sustainability)
UW	University of Washington
V or Vol	volume, vacuum, volt (electrical), vent, valve, or vertical
VE	value engineering
VEA	value engineering analysis
VEM	value engineering method
VP	vice president
W	west, width, wide flange (steel beam), waste, water, watt (electrical), or washer (clothes)
W or w/	with

WA	the state of Washington
WBS	work breakdown structure
WF	wide flange (steel beam or column); formerly 'I' or 'H' beams
WK or wks	weeks
w/o	without
WRT	with respect to
WT	weight
WWF	welded wire fabric, also known as wire mesh (concrete reinforcement)
×	times (multiplication), cross bracing, or "by," as in dimensional lumber (for example: 2×4)
YD	yard
YR	year

Part I

Introductory Topics

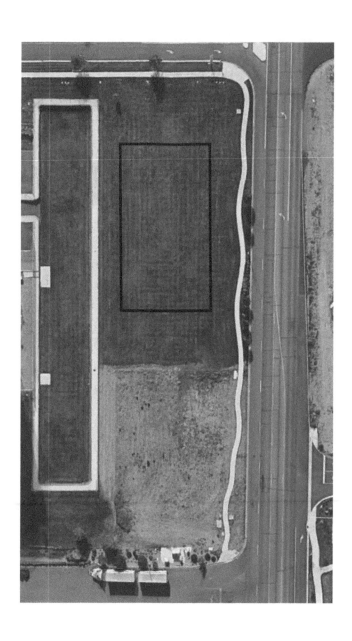

Part 1

Introductory Topics

Introduction

Like estimating, construction scheduling has been covered in other books dedicated solely to that topic. But scheduling by itself, without integrating preconstruction planning and controls during construction, is an incomplete construction management (CM) study. The schedule is a project management (PM) tool just as is the estimate; and time management is just as important to project success as is cost management. The key to effective time management is to carefully *plan* the work to be performed, develop a realistic construction *schedule*, and then *control* or manage the performance of the work. People often use the terms *planning* and *scheduling* together. Planning is the up-front work which makes the schedule feasible. Planning is a process and the schedule is the result. The schedule is a logical arrangement of activities in order of occurrence, with prerequisites, and charted with a timeline. And the control element is the implementation of those plans during physical construction of a project. Planning, scheduling, and control are therefore the three major sections and focus of this book.

Schedules are important tools for all members of the owner and design and construction teams. Proper planning of the project and the schedule, with input from the relevant personnel such as the general contractor's (GC's) project superintendent and major subcontractors, are keys to developing a useful construction management tool. Schedule development begins with proper planning, which considers many variables such as deliveries, logical workflow, manpower, and equipment availabilities. There are many different types of schedules, each of which has a use on a construction project. Some of the major ones include:

- Summary schedule, may also be the contract schedule;
- Detailed schedule, may be the contract schedule;
- Three-week look-ahead schedules;
- Specialty schedules which include those focused on one area of the building or phase or on just one subcontractor;
- Pull planning schedules, which are part of lean construction;
- Expediting and submittal schedules and others.

1.1 SCHEDULERS

Who draws schedules and who uses schedules? Hopefully after reading the book you will have a thorough understanding of this topic. This chapter provides just a brief introduction.

First, some members of the built environment use the terms *scheduling* and *project management* as though they are the same, but PM as presented in this book involves much more than scheduling. A PM may be the scheduler on his or her particular project, but a person who schedules by profession, be they a home office specialist or a scheduling consultant, would not typically also be a project manager. For many the "scheduler" is the one in charge of creating the schedule, such as a superintendent or project manager, but for others the scheduler is a computer technician, sometimes without extensive construction field experience. A valuable scheduler oftentimes is one with a mix of construction field knowledge and technical skills.

Different contractors will establish the role of the scheduler in a variety of fashions. A scheduler can be assigned to the home office as a staff scheduler and he or she either prepares all the project schedules (for a small to mid-sized contractor) or supports the project manager and superintendent with creating their own schedules. Or a large project may have its own full-time scheduler. If there is not someone in-house, then an outside consultant/specialist may be hired, or a project engineer or assistant superintendent may be the project scheduler. A retired superintendent working as a consultant, teamed with a construction management graduate, is an excellent example of a scheduler as shown in this first example. The problem with either the PM or superintendent also taking full responsibility for the schedule maintenance on a large project is that the effort may be all-consuming, as will be shown in another example further on in the book. An industry partner from a national construction firm who was interviewed when researching this book indicated, "The scheduling process requires teamwork and does not rely on an isolated home office individual. The ideal scheduler must have a background in construction, particularly with field expertise."

Example 1.1

One of the area's most highly acclaimed superintendents retired, but he wanted to keep himself busy and had always enjoyed drawing schedules by hand. Another competing GC hired him as their scheduling vice president and he would come into the office one or two days a week and assist field superintendents with their schedules. He would review the drawings first, then sit down with the field superintendent and a large sheet of butcher paper and scratch the schedule out, with lots of loop lines and plenty of erasures, but somehow it all got down on paper. The field superintendents were happy to work with the scheduler as they all appreciated his experience and insights. He would have a young CM graduate formalize the schedule and transform it into a useful communication tool. During the course of construction, he was on call if a superintendent needed support, but he generally left jobsite controls, including schedule control, to the field supervisors.

1.2   SCHEDULE TYPES

Schedules take on a lot of different formats, and similar to planning, there is no one exact form the contract schedule should follow. Many project managers and schedulers have their personal preferences. Most schedules fall into one of two standard formats: Bar charts and network diagrams. Bar charts relate activities to a calendar, but generally show little to no relationship among the activities. Network diagrams show the relationship among the activities, and may or may not be time-scaled on a calendar. Two diagramming techniques are used to represent network schedules. The first is known as the arrow diagramming method (ADM), in which arrows connect the individual activities and nodes depict events. The other is known as the precedence diagramming method (PDM), in which the activities are represented by nodes and arrows depict relationships between activities. Arrow diagramming method was the original form of *critical path* schedules but PDM has basically replaced ADM today. Network diagram schedules are also customarily referred to as critical path schedules. The critical path is the longest path through the schedule and determines the overall project duration. Any delay in any activity on the critical path results in a delay in the completion of the project. All are good systems and may be appropriate in different applications.

The schedule is a tool to help build the construction project and is used in a variety of fashions, as exhibited in this second example. Schedules can also be prepared for different presentations depending upon the anticipated use. Detailed discussions of all the following schedule types are elaborated on in Chapter 7. There are many other types of schedules and scheduling theories than presented in this book, each with an entirely separate glossary of terms and lists of abbreviations. A few of these other scheduling concepts are introduced in this book as well, but a more detailed analysis will be left for a more advanced book geared for the professional schedule consultant. This focus is on the scheduling tools typically utilized in the field by construction contractors. Many of these useful scheduling tools are described here.

Example 1.2

Schedules are intended to be tools to help build a project. A developer, who was also his own contractor, was a very capable builder and creative but did not have the patience or interest to draw formal schedules. Instead he used two whiteboards in the jobsite trailer – one for this month and one for the next month. This was an effective communication tool for the foremen and subcontractors on his current apartment project. But this type of schedule was not sufficient for the bank or his investment partners. The developer hired a scheduling consultant who prepared good detailed schedules to present to the bank each month with the monthly pay request draw, but he never hung them in the trailer nor shared them with subcontractors or the design team. This schedule was also a tool, but not the tool that it might have been, given more attention.

- *Contract schedules*, also known as formal schedules: These schedules will be provided to the client at the beginning and throughout the project delivery as required by the contract special conditions.

- *Summary schedules*: These schedules are often used for presentations or proposals or management reporting. They are similar to the milestone bar chart illustrated in Figure 1.1 for the case study project. A more detailed summary schedule is included in Chapter 7.

- *Detailed schedules*: These schedules are posted on the walls of meeting rooms or in the jobsite trailer, or at least they used to be. Now many are kept electronically on the computer screen as discussed in a couple of boxed-in examples presented further on in the book. These schedules are marked up with comments and progress. Detailed schedules may also serve as submittals for contract schedule requirements. A complete detailed schedule for the case study project is included on the book's companion website. Portions of this detailed schedule are used as figures and examples throughout the book.

- *Short-interval look-ahead schedules*: These schedules focus on short-term field activities. They should be developed by each superintendent or foreman and each subcontractor each week. These schedules can be in two-, three-, or four-week increments, depending on the job and level of activity. Some contractors simply print the next three weeks' worth of activities from the electronic master schedule without any additional detail or input from the superintendent, which is not as effective. Short-interval schedules may also be referred to as trade schedules in that each subcontractor prepares their own. These schedules are valuable for pretask planning, similar to pull schedules discussed further on. Figure 1.2 shows an example of the initial three-week schedule from the case study project and additional examples are included throughout the book. The schedule format is not as important as its author and content.

- *Mini-schedules*, *area schedules*, and *system schedules*: These schedules are intended to allow additional detail for certain portions of the work that could not be adequately represented in the detailed project schedule and have longer durations than the short-interval schedule. A system schedule for the case study's roofing subcontractor's scope is also included on the companion website.

- *Pull schedules*: These schedules are one of the major lean construction tools being adapted from production industries. They are prepared by the *last planners*, typically the foremen and superintendents responsible for accomplishing the work. Pull planning schedules are discussed in lieu of schedules in Chapter 5.

- *Other schedules*: There may be many other specialized schedules on the project that include submittal, buyout or procurement, delivery, start-up, as-built, and close-out schedules. Estimate or bid schedules are utilized to prepare the original quote and may be included with the bid. Many of these are discussed in other chapters in this book.

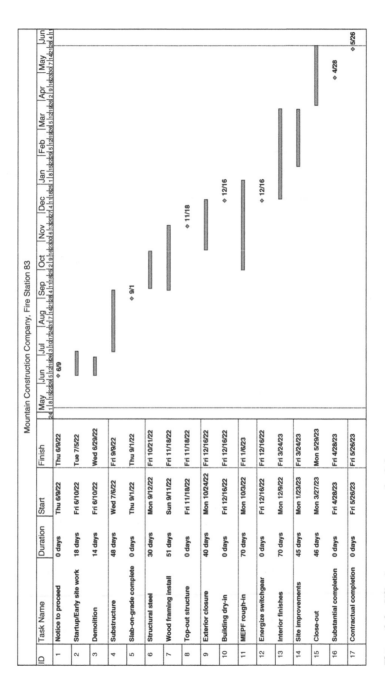

Figure 1.1   Milestone summary schedule

Mountain Construction Co.
Short-interval schedule

Project: Fire Station 83, Pasco, WA Date: 6/6/2022
Superintendent: Ralph Henry Sheet: 1 of 1

June

No.	Activity description:	S	S	6	7	8	9	10	S	S	13	14	15	16	17	S	S	20	21	22	23	24	S	S	Comments
1	Notice to proceed						X																		Original in office
2	Deliver office trailers							X																	
3	Install silt fence							X			X	X													
4	Surveyor:			Site and bldg corners							Layout utilities							Building gridlines/setbacks							
5	Locate utilities							X			X	X													Public utility co.
6	Install temp power												X	X	X			X	X						
7	Deliver sanikan							X																	
8	Clear and grub site												X	X											
9	Pre-con meeting with city											X													
10	Mobilize demolition sub											X													Demo permit ready?
11	Demo existing structure												X	X											
12	Geo approve import sample										X	X	X	X	X			X	X						Expedite!
13	Site utilities																	X	X	X	X	>			Post demo
14	Rebar shop drawings										X	X	X	X	X			X	X	X	X				Expedite!
15	Deliver rebar																					X			
16	Structural excavation																					>			Tues, 6/28
17	Begin foundations																					>			6-Jul

Figure 1.2 Three-week schedule

1.3 INTRODUCTION TO THE BOOK

Construction has its own unique set of abbreviations and acronyms. An extensive list of all of the abbreviations used in this book, and several other industry standards, is included at the front of the book. Over 80 figures and tables that help to explain concepts are included in the book. Many of these are connected to the book's primary case study, as discussed further on in this chapter. Lists of these illustrations are also included in the front material.

The book is organized into four major parts, beginning with this first part, "Introductory Topics." In addition to this chapter. which defines the scheduler and introduces several types of schedules, all of which will be expanded on throughout the book, this first chapter outlines the layout of the book and introduces the reader to the case study. The second chapter includes the many introductory topics that form the foundation for any CM discussion, especially scheduling.

Part II, "Planning," explains the creative part of scheduling, the *planning* phase. Many preconstruction plans are developed by a contractor before construction commences. Planning involves breaking the work down into identifiable and measurable work tasks, often referred to as work packages. These tasks, or activities, are then arranged in a logical order, resolving the question "What precedes what?" How one activity relates to another is known, in scheduling terms, as relationships or restraints, and these are incorporated into the logic plan. Other planning considerations include sustainability, lean practices, and a thorough review of proposed contract language.

Very few contractors have professional "planners." This upfront work may be performed by a scheduling expert, but input from the builders, such as the PM and superintendent, are vital to the preparation of a successful plan. Without a good plan, schedules are rarely viable construction tools, as reflected in this quote from a former contractor-employer:

> *We never have time to build it right,*
> *But we always seem to have time to build it over.*

Scheduling is the second phase, which involves adding durations to activities and a timeframe and milestone dates, or constraint dates to the logic diagram and plotting out a schedule document. Many calculations are performed today by scheduling software that highlights the *critical path* of the project, or those activities that must be completed either on time or earlier or they will cause delays for the entire project. Activities which are not deemed to be critical have float. The scheduling effort encompasses many topics that are included in Part III, "Scheduling," including resource planning and technology tools. As discussed previously, the person who transformed the plan into the schedule is the scheduler and that role may be performed by a home office specialist or a field supervisor.

After the contractor is done preparing the plan and has incorporated it into a schedule document, then construction commences and it is time to make sure that the work is performed according to the plan. Some may call this "monitoring" or "recording" but *controlling* the schedule is the common nomenclature and Part IV, "Project Controls," includes many construction management control topics and tools. In addition to schedule control, this fourth section of the book includes earned value management, subcontract management, and schedule impacts, including construction claims. The actual progress of the work is measured and compared against the schedule. If there are deviations, then an adjustment to the plan is necessary which often results in a revised schedule. There are many members of the jobsite team involved with controls, including the project manager, superintendent, cost and schedule engineers, foremen, and project engineers.

Figure 1.3 is a flowchart of many of the planning, scheduling, and control activities that are discussed throughout this book. There are other subsequent flowcharts that are subsets and expand on various portions of this initial chart. Essentially Figure 1.3 is a plan for the book. A full-size copy of this figure is included on the book's website and the reader may wish to print out a copy of that flowchart to use as their bookmark.

Just as construction documents (and books) are full of abbreviations, there are also unique words and terms in the built environment industry. An extensive glossary of scheduling and construction management terms encompassing all of those introduced in the body of this book, as well as many other common built environment terms, are included as an appendix. Several other publications were drawn on while researching these topics, many of which are referenced in the book, and a list of these books is included as an appendix that the scheduling student or construction professional might find useful for their personal library.

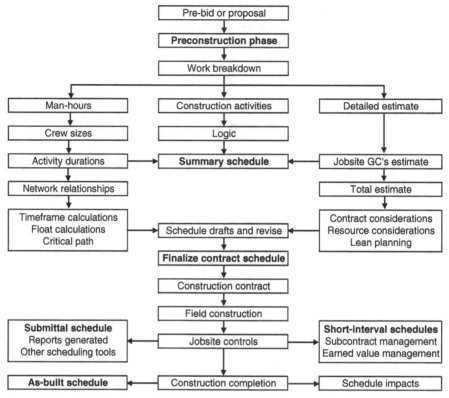

Note: Additional detail is added in Flowcharts 4.1, 9.1, and 14.1.

Figure 1.3 Planning, scheduling, and control flowchart

As discussed next, the book has one substantial project that is threaded throughout many of the figures and discussions. Over 35 other examples from actual construction projects are set in boxed features and discussed in the book; two of them were included in this chapter. These boxed examples connect with the relative chapter topic and are from the author's firsthand experiences, some of them successful and some not so, but all offer good learning opportunities. In addition, five short case studies have been borrowed from *101 Case Studies in Construction Management* that are directly connected with scheduling and they are also included at the back of the book as an appendix. These short cases are presented in a different format than the book's primary case study. The five *101 Case Studies* bring up problems that have occurred on various projects and lead the reader into a "Who done it?" scenario without any one exact correct answer.

1.4 INTRODUCTION TO THE CASE STUDY

The Fire Station 83 Project is a new single-story fire station constructed in Pasco, Washington. The building comprises 10,612 square feet and consists of a variety of spaces including: Four drive-through vehicle bays, sleeping rooms and living space for the crew, an exercise facility, kitchen, offices, and several areas to support fire-fighting operations, including cleaning and decontamination rooms, mechanical and electrical rooms, and a vehicle maintenance shop.

The building's structure is a combination of wood and steel framing supported by cast-in-place concrete spread and continuous footings with a concrete slab-on-grade. Important structural components include a thickened eight-inch thick slab-on-grade throughout the vehicle bay and 22-foot tall structural steel columns to support the vehicle bay's steel roof joists. Exterior enclosure components include metal wall, roof, and soffit panels; aluminum-framed storefront glass; fiberglass and vision-control glazing; and phenolic wood panels. Construction of the site includes clear and grub, excavation, permanent site utilities, concrete and asphalt pavement, curbs and sidewalks, site structures, and landscaping.

A rendering of the building, which was included with the bid documents, is shown in Figure 1.4. The building's owner is the City of Pasco and the architect was TCA Architecture, Planning, and Design of Seattle, Washington. We appreciate Pasco and TCA for allowing the use of their project as the book's primary case study. The project was competitively bid at a little over $4.7 million and took just under a year to build.

The drawings and specifications for the actual project are included on the companion website. A list of all of the website support documents is included in the book's front material. Several examples of schedules and estimates and other background information are threaded throughout the book as figures and tables. These documents connect the major topics of planning, scheduling, and construction management controls together. Many of the applied exercises in the book are also related

Photo courtesy of TCA Architecture, Planning, and Design

Figure 1.4 Case study rendering

to this case study. Additional detailed backup including a complete schedule, detailed estimate, and site logistics plan are also included on the companion website. Although this is a real project that is used in this book as a great example to discuss scheduling, all of the figures and tables in the book have been re-created for educational purposes. The fictional general contractor utilized for the case study in this book is Mountain Construction Company (MCC). A traditional GC organization was assumed for this project. Any connection with actual companies or individuals who participated in the construction project is coincidental. It is also assumed the preconstruction manager working for Mountain Construction prepared the bid estimate and detailed and summary schedules, and then transitioned into the role of jobsite PM for the duration of construction. A few examples for an additional negotiated project are also included as figures in the book.

1.5 SUMMARY

This is not a book about how to become a professional scheduling consultant; rather the focus is on developing planning, scheduling, and control tools for the construction management student and the industry professional's (project manager and superintendent's) tool box. Planning is a proposal to act. Scheduling involves creation of a written document to carry out the plan, essentially adding timeframes and durations to the original plan. Controlling is monitoring the schedule and making adjustments as necessary so that the original plan is accomplished. The importance of proper planning is critical to production of a schedule communication tool that can be managed in the field.

Many people involved in real estate development, including designers and builders, prepare schedules. The book's focus is on the scheduler for the construction contractor. This person may be a home-office staff specialist, outside consultant, or field supervisor. There are many different types of construction schedules and scheduling tools, from summary to detailed schedules and from short-interval foreman's schedules to as-built schedules. All of these tools are discussed in this book. The book's primary case study project is a $4.7 million fire station built in Pasco, Washington. The construction duration was 11 months. Many of the figures and exercises presented in the book connect with that project. In addition to the commercial fire station case study, schedule examples for residential and heavy civil projects and a specialty contractor schedule are also included on the companion website.

1.6 REVIEW QUESTIONS

1. Have you ever worked as a scheduler or with a scheduler?
2. Looking at the list of types of people or companies presented earlier who might be schedulers, which would be the best fit for these situations:

A. Small GC that does residential remodel work

B. Mid-size GC that builds work in the $5 to $50 million range

C. National CM with 1,000 management employees

D. Public utility company

E. Attorney who represents contractors in legal disputes

3. Why did the field supervisors trust the scheduler in Example 1.1 maybe more than they would an outside consultant? There are a couple of reasons.

4. Was either of the schedules described in Example 1.2 an effective construction management tool? Could one of them have replaced the other? Could a third schedule have replaced them both?

Chapter **2**

Construction Management

2.1 INTRODUCTION

This chapter has been included in the scheduling book, along with other introductory topics, to provide a brief overview and introduction to current construction management (CM) processes. It is important to introduce the reader to a few basic construction management terms and processes before diving deeper into planning, scheduling, and controlling construction projects. This chapter is not a stand-alone treatise on the very broad and important topic of CM; CM deserves its own book or separate class on just that area. The focus here is on the relationship between project management (PM) and scheduling, and other construction management topics, as they relate to jobsite planning, scheduling, and control. For a more thorough coverage of CM and PM the reader is suggested to look to a more comprehensive resource such as *Management of Construction Projects, A Contractor's Perspective*. Much of the material in this chapter has relied upon that source.

The focus of construction management here is on the role of the project manager as an individual and not an outside company. Many concepts and terms from one (CM or PM) apply to the other. The CM and the PM are both builders, but so are many other members of the built environment team. The term *builder* is too generic—everyone on the project, including the owner, architect, general contractor (GC), and the craftsmen contribute to the building process and each is a "builder," per se. There are also many additional levels of specializations within the GC's organization as discussed later. There are many other BE participants, including specialty engineers and consultants, many of which are listed here:

- Structural engineer
- Civil engineer
- Mechanical engineer
- Electrical engineer
- Kitchen consultant
- Landscape architect
- Interior designer
- Elevator consultant

- Accountant
- Banker
- City inspectors
- Third-party inspectors
- Commissioning agent
- Envelope/waterproofing consultant

This chapter will describe the delivery and procurement methods the project owner chooses to use as well as contracting and pricing options that the contractor must incorporate into the preconstruction plan. The type of cost estimate prepared by the contractor and its degree of accuracy is often governed by the owner's choice of delivery and procurement methods. Different types of general construction organizations are also introduced as well as the roles of project managers and schedulers and the risks they must evaluate when considering a potential construction project.

2.2 DELIVERY AND PROCUREMENT METHODS

The three major companies and/or individuals that are the primary responsible parties in any construction project include the project owner or client, the designer (architect or engineer), and the GC or CM. The relationships among these participants are defined by the procurement and delivery methods used for the project. The choice of delivery method is the owner's, but it has an impact on the responsibilities of the contractor's jobsite team. The project owner selects a delivery method based on their experience and the risk they are willing to absorb or pass to either the GC or the design team. This section examines the four most common delivery methods. Each of these different arrangements influence how the GC will conduct its jobsite project management operations, including scheduling. Basic differences in bid and negotiated procurement processes are also introduced.

Traditional General Contractor Delivery

The most common method of project delivery is the traditional method and is represented in the organization chart Figure 2.1, position 1. The owner has separate contracts with both the designer and the general contractor. There is no contractual relationship between the designer and the general contractor. Typically, the design is completed before the contractor is hired in this delivery method. The GC's project manager takes the point to obtain the project drawings and specifications, develop a cost estimate and construction schedule, establish control systems to manage construction activities, and manage the construction site office. A traditional GC employs a mix of subcontractors and direct craftsmen. The case study and exercises used throughout this book is based upon the traditional GC delivery method.

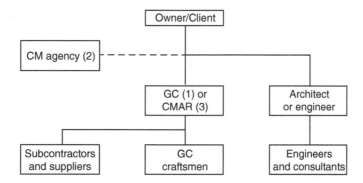

Figure 2.1 General contractor and construction manager delivery methods organization charts

Construction Management Delivery

There are two basic construction management delivery methods. One is the agency CM and the other is the at-risk CM. The agency CM does not employ any subcontractors or direct craftsmen and the pure at-risk CM employs all subcontractors but no direct craftsmen, different from the mix employed by the traditional GC as described above.

The project owner has three separate contracts (one with the designer, one with the general contractor, and one with the construction manager) in the *agency construction management* delivery method as shown in Figure 2.1, position 2. The construction manager acts as the owner's agent and coordinates design and construction issues with the designer and the GC. The construction manager usually is the first contract awarded, and he or she is involved in selecting both the designer and the GC. In this delivery method, the general contractor usually is not hired until the design is completed. The Agency CM is sometimes referred to as the owner's representative and typically does not have any financial risk.

In the *construction manager at-risk* (CMAR) delivery method, the owner has two contracts (one with the designer and one with the construction manager) as illustrated in Figure 2.1, position 3. This delivery method is also known as the *construction manager/general contractor* (CM/GC) delivery method. In this case, the designer usually is hired first, but the CM/GC is also contracted early in the design development to perform a variety of preconstruction services, such as cost estimating, constructability analysis, and scheduling. Once the design is completed, the CM/GC constructs the project. In some cases, construction may be initiated before the entire design is completed. This is known as fast-track or phased construction. One additional difference between a CMAR and a traditional GC is that the GC customarily employs some craftsmen directly, such as carpenters and laborers, for concrete work, but the CMAR typically subcontracts out 100% of the construction labor (in fact, this is often a contract requirement).

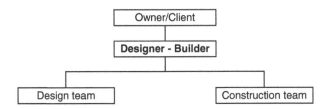

Figure 2.2   Design-build delivery method organization chart

Design-Build Delivery

The project owner has a single contract with a design-build (DB) contractor for both the design and construction of the project in the design-build delivery method, as diagrammed in Figure 2.2. The DB contractor may have a design capability within its own organization, may choose to enter into a joint venture with a design firm, or may hire a design firm to develop the design. The scheduler working for a DB contractor will by definition include design activities within the construction schedule. Construction may be initiated early in the design process using fast-track procedures or may wait until the design is completed.

There are a variety of other delivery methods, some of which are hybrids of these four. The *integrated project delivery* (IPD) is a relatively new concept in which all three prime parties sign the same agreement and share in the same risks. A sample IPD organization chart is included in Chapter 5, "Lean Construction Planning." In the five primes or *multiple-prime delivery* method, the project owner contracts with several general contractors or major subcontractors at the same time. This often includes a site work contractor, building shell and core contractor, tenant-improvement contractor, mechanical contractor, electrical contractor, and/or others. The multiple-prime delivery method usually requires a construction-experienced project owner to act as their own CM. Another alternate delivery process is known as *public-private partnership*, abbreviated as PPP or P3. This process combines the resources of a private entity such as a real estate developer that may own a piece of property, with the needs of a public client, such as a university looking to build a research laboratory. The private developer will design, finance, build, and operate the facility for benefit of the university, which will sign a long-term lease guaranteeing the developer's pro forma will work out. The public client does not need to go through the sometimes complicated and often litigious lump sum bid process when utilizing the P3 delivery method.

Procurement

Project owners solicit or procure both construction and design team members utilizing either bid or negotiated procedures. Public project owners are often required by law to use an open bidding process to allow all contractors an even chance of successfully landing new work. Private owners can use whatever procurement method they choose, bid or negotiated, but often solicit contractors they have had good experiences with in the past, and may ask a select few or even only one prequalified contractor to submit a bid or negotiated proposal. Major subcontractors and suppliers can also be prequalified and vetted, which helps reduce the risk for the bidding GCs. There are risks and advantages associated with either bidding or negotiating for the owner, architect, and contractor. This book will utilize a contractor chosen for a public project that was competitively bid as an example case study and as the basis for many of the exercises.

2.3 CONTRACTS

The prime contract is the agreement between the general contractor or construction manager and the project owner. It is a legal document that describes the rights and responsibilities of the parties. The focus here is on the prime contract agreement, but many of the documents and processes described apply to subcontract agreements and supplier purchase orders as well. The terms and conditions of the relationship of the contracting parties are defined solely within the contract documents. These documents should be read and completely understood by the contractor before deciding to pursue a project and prepare a bid. They also are the basis for determining a project budget and schedule. To manage a project successfully, the GC's PM must understand the organization of the contract documents and contractual requirements, including schedule requirements, for his or her project.

The contract documents describe the completed project and the terms and conditions of the contractual relationship between the owner and the contractor. Usually there is no description of the sequence of work or the means and methods to be used by the contractor to complete the project. The contractor is expected to have the professional expertise required to understand the contract documents and select appropriate subcontractors or qualified tradespeople, materials, and equipment to complete the project safely and achieve the quality requirements specified. For example, the contract documents will specify the dimensions and workmanship requirements for elevated concrete slabs but will not provide the design for required formwork or methods for temporary shoring or reshoring. The contract usually includes the agreement itself, both special and general conditions, drawings, and technical specifications and all these documents must work together. A figure reflecting the relationship of the contract documents is included in a later chapter.

Project owners typically use contract formats developed by the American Institute of Architects (AIA) or the new family of construction contracts from ConsensusDocs headed up by the Association of General Contractors. Very large institutional owners may have specialized agreements drawn up by their legal counsel. Contracts should not be signed until they have been subjected to a thorough legal review. The contract documents have a significant impact on the responsibilities of the PM and superintendent. Many requirements are contained in the general conditions, but project-specific requirements, such as work days and hours, are defined in the special conditions of the contract. Many contract issues, such as liquidated damages and claim notifications, are related to how contractors prepare and control the construction schedule, and a complete chapter, Chapter 6, has been dedicated to just those topics.

2.4 PRICING

There are several methods for pricing contracts used in the construction industry but four are the most prevalent and others are hybrids of these four. The choice of which to use is also made by the owner after analyzing the risk associated with the project and deciding how much of the risk to assume and how much to pass on to the contractor. Contractors want compensation for risk they assume and usually do so with increased fees or estimating contingencies. The most common pricing methods include:

- Lump sum (LS),
- Unit price,
- Cost-plus fixed fee or cost-plus percentage fee, and
- Cost-plus with a guaranteed maximum price (GMP).

Lump sum, or stipulated sum contracts are awarded on the basis of a single lump sum estimate for a specified scope of work. Unit-price contracts are utilized for heavy civil projects when the exact quantities of work cannot be defined. Cost-plus contracts are used when the complete scope of work cannot be defined or construction needs to start before design is complete. All the contractor's project-related costs are reimbursed by the owner, and a fee is paid to cover profit and the contractor's home office overhead. This may also be known as a time and materials contract. A GMP contract is an open-book cost-plus contract in which the contractor agrees not to exceed a specified cost.

Project owners choose which method they want their contractors to price a project for a variety of reasons, including the completeness of the design, the complexity of the project, and the owner's experience in managing construction projects. Each of these pricing models also has a significantly different effect on jobsite management, including scheduling. The LS pricing method is utilized for the case study and examples in this book and it will introduce the reader to a variety of cost and scheduling opportunities. These four pricing options are also reinforced in the next chapter on preconstruction.

2.5 ESTIMATING

Cost is one of the most critical project attributes that must be controlled by the project management team. Project costs are estimated to develop a construction budget within which the project team must build the project. Cost estimating is the process of collecting, analyzing, and summarizing data in order to prepare an educated projection of the anticipated cost of a project.

Project cost estimates may be prepared either by the project manager or by the estimating department of the construction firm. Similar to scheduling, when possible, the PM and superintendent should be responsible for developing the estimate, or at a minimum work as integral members of the estimating team. Their individual inputs regarding constructability and their personal commitments to the final estimate are essential to assure not only the success of the estimate, but also the ultimate success of the project.

Many outside of the construction industry view all contractors' estimates the same, in that they are all "firm bids" and all contractor-produced estimates are completely detailed and accurate. This of course is a misconception and largely dependent upon document completion. There are several different types of cost estimates. Conceptual cost estimates are developed using incomplete project documentation, while detailed cost estimates are prepared using complete drawings and specifications. Semi-detailed cost estimates are used for GMP contracts and have elements of both conceptual and detailed estimates. The estimating strategy or approach is different with each of the three main types of estimates and the level of detail will differ as well. The accuracy of an estimate is directly proportional to the accuracy of the documents and the time spent on preparing an estimate also follows those same lines. Schedule preparation follows a similar process. All estimates have major elements or cost categories, some of which require a more detailed effort by the estimator than others.

Early estimates may be developed by contractors or architects on limited information and produced quite quickly. These estimates are not expected to be "firm" nor are they necessarily accurate. They should be referred to as *budgets, schematic estimates,* or *conceptual estimates.* A conceptual set of drawings can be estimated quite quickly using square foot of floor (SFF) unit prices, assembly prices, subcontractor plugs or budgets, and percentage add-ons for general conditions to produce a rough-order-of-magnitude (ROM) budget. Completion of the schematic design phase is also usually followed by a contractor-generated, or estimating consultant-generated, budget estimate, but this is not a firm bid. All of the five different design phases are described more fully in the next chapter on preconstruction. Most items within a budget are by definition allowances or plugs. Budgets are the least accurate estimate type and should carry substantial contingencies, such as 10–20%. Budgets are also developed only by seasoned estimators, whereas the junior estimator or cost engineer will assist with quantity take-offs associated with detailed LS bid estimates.

A contractor will also develop an early and quick budget from even a detailed set of drawings as soon as they come through the door. This estimate is referred to as a ROM estimate and is utilized to determine if the project is the right size for the contractor to pursue and which of its estimating or management staff would be best suited to work on the subsequent detailed estimate.

A *detailed estimate* takes the longest time to prepare, costs the contractor the most in personnel resources to complete, and produces the most accurate final figure. Usually drawings and specifications that are 90–95% complete are associated with detailed estimates. This would be consistent with completion of the construction document design phase. Detailed estimates are required for projects which are lump sum bid, such as for a public waste water treatment plant or a fire station.

Although lump sum bids are customarily associated with public works projects, clients on privately financed projects may also solicit lump sum bids during a slow economy. In a busy economy, though, project owners may only interest contractors with negotiated requests for proposals. Clients soliciting lump sum bids are primarily interested in just the bottom-line price and assume all contractors can deliver the project with comparable levels of quality, schedule, and safety. Mountain Construction Company provided a LS bid estimate and entered into an AIA A101 contract with the City of Pasco for the Fire Station 83 case study project referred to throughout this book.

Guaranteed maximum price estimates and resultant guaranteed maximum price contracts are a hybrid of budgets and detailed estimates. Guaranteed maximum prices are often prepared on private negotiated projects after completion of the design development phase. Detailed estimates are produced for scopes which have been adequately designed and specified, such as structural concrete and steel, utilizing measured quantities and unit prices. Subcontractor plugs or allowances are included for areas not yet designed, but competitive subcontractor bids are factored wherever possible. For work which is not fully designed, the contractor will use assembly costs such as $/SFF or allowances. Contingencies are more prevalent in GMP estimates than detailed estimates and may amount to 2–5%. Projects with a GMP are usually performed open-book and any resultant savings are shared between the client and general contractor, such as 80% to the client and 20% to the GC.

Similar to constructing a building, estimating is a logical process consisting of a series of steps, the first being project overview to determine if the project is going to be pursued. Once decided, a work breakdown structure should be outlined which will assist with both the estimate and construction schedule development. The quantity take-off step is a compilation of counting items and measuring volumes. Pricing is divided between direct labor, materials, construction equipment, and subcontracts. Labor cost is computed using productivity rates and current local labor wage rates. Material, equipment, and subcontract prices are developed most accurately using competitively bid supplier and subcontractor quotations. Jobsite general conditions cost is a job cost and is schedule dependent. Home office overhead is combined with desired profit to produce the fee percentage. The fee calculation on any specific project varies dependent upon several conditions including company volume, market conditions, labor risk, and resource allocations. All estimates, especially detailed estimates, are composed of these basic elements and as shown in the equation below. These major estimate areas are also shown in the case study's summary estimate, Figure 2.3. A complete detailed estimate for this project, along with a detailed jobsite general conditions estimate, is included on the companion website.

Mountain Construction Company
Fire Station 83, Project #9922

Summary estimate   11 months   10,612 SF

CSI Div	Description	Direct L. HRs (2)	Labor	Mat'l/Equip	Subs	Total
01	Jobsite general conditions (1)		$271,410	$121,955		$393,365
02	Demolition				$94,409	$94,409
03	Concrete	1,525	$72,805	$124,755		$197,560
04	Masonry				$0	$0
05	Structural & misc. steel	1,482	$93,475	$277,486		$370,961
06	Rough and finish carpentry	1,193	$56,533	$124,764	$93,794	$275,091
07	Thermal and moisture				$356,834	$356,834
08	Doors and windows	369	$17,488	$78,051	$209,393	$304,932
09	Finishes				$147,273	$147,273
10	Specialties	102	$5,235	$29,822	$3,500	$38,557
11	Equipment	28	$1,326			$1,326
12/3	Furnishings				$9,047	$9,047
14	Elevators				$0	$0
21	Fire protection				$75,000	$75,000
22	Plumbing				$200,000	$200,000
23	HVAC and controls				$467,300	$467,300
26/7	Electrical				$825,727	$825,727
31	Earthwork				$135,824	$135,824
32	Exterior improvements	1,259	$59,375	$123,077	$127,605	$310,057
33	Offsite improvements	163	$8,669	$14,432	$7,558	$30,659
	Subtotals:	6,121	$586,316	$894,342	$2,753,264	$4,233,922

Labor burden on direct labor		$115,168	$4,349,090
Labor burden on general conditions		$70,847	$4,419,937
Subcontractor bonds	w/sub $	$0	$4,419,937
State excise, B & O tax	0.48%	$21,392	$4,441,329
Liability insurance	1.1%	$48,855	$4,490,184
Builder's risk insurance	by owner	$0	$4,490,184
Fee	5%	$226,754	$4,716,938
Contingency	0.00%	$0	$4,716,938
Total base bid:			$4,716,938
GC P & P bond	0.95%	$44,903	$44,903
Total bid with P & P bond:			$4,761,841

Notes: 1) See separate detailed estimate $448.72
        2) GC hours are for direct labor only $/SFF

Figure 2.3 Case study summary estimate

Direct labor

+Direct material

+Construction equipment

+Subcontractor costs

+Jobsite general conditions

+Markups including fee
_____

= Construction cost estimate

2.6 PROJECT MANAGEMENT

Construction project success is generally defined in terms of safety, quality, cost, schedule, and document control. The construction team's challenge is to balance quality, cost, and schedule within the context of a safe project environment and document all aspects accordingly. Safety is of utmost importance to companies and individuals in the built environment industry, and sacrificing any aspect of safety to improve quality, schedule, or cost performance is unethical, dangerous, and unacceptable to all participants.

Most of the book approaches project management and scheduling from the perspective of the general construction contractor. Other project managers typically are involved in a project representing the owner and the designer and most of the principles presented here apply to them as well. The context of this book will be that of project managers and schedulers working for mid-sized commercial general contractors. The principles and techniques discussed, however, are equally applicable on residential, industrial, and infrastructure or heavy civil construction projects, as well as for specialty subcontractors.

The project manager and superintendent share leadership oversight of the contractor's jobsite project team and are responsible for identifying project requirements and leading the team in ensuring that all are accomplished safely and within the desired budget and timeframe. The focus of the superintendent is on the field installation side and the PM on the office and field management side. To accomplish this challenging task, the project manager must organize his or her project team, establish project management and cost and schedule control systems that monitor project execution, and resolve issues that arise during construction.

There are typically four major phases of a construction project. These phases include planning or preconstruction, startup, control, and close-out. Some may add a fifth post-project analysis phase although this is not always practiced or it is included with close-out. During project *planning*, the project manager, superintendent, and upper management evaluate specific risks that are associated with the project, particularly those related to safety, cost, quality, and schedule. The contractor develops the organizational structure needed to manage the project. Material procurement and subcontracting strategies are also developed during the planning phase, and along with jobsite material management, will be discussed in Chapter 17. The jobsite office is established during *start-up* and control systems are initiated. Vendor accounts are established, and materials and subcontract procurement initiated. Project cost, schedule, safety, quality, and document control systems are established to manage all aspects of project execution.

Project control is a broad term that involves controlling or managing the project during construction, interfacing with external members of the project team, anticipating risks by taking measures to mitigate potential impacts, and adjusting the project schedule to accommodate changed conditions. The project manager and superintendent monitor the document management system, quality management, cost control, and schedule control systems, making adjustments where appropriate. Many performance reporting tools are implemented to look for variances from the original plan and action is taken to minimize impacts to overall project success. *Project close-out* includes not only completion of the physical construction of the project, but also submission of all required documentation to the owner, and financial close-out. The project manager must pay close attention to detail and motivate the project team to close out the project expeditiously to minimize jobsite overhead costs. One in-house close-out task is development of an as-built schedule. All of these control aspects are included in Part IV of this book.

2.7 GENERAL CONTRACTOR ORGANIZATIONS

The terms *builders, constructors,* or *contractors* include not only general contractors and construction managers, but specialty or subcontractors as well. General contractors are differentiated from subcontractors where applicable in this book, but the generic term *contractors* is also used to include all built environment participants who contribute direct labor on the jobsite and develop and utilize construction schedules. The size and structure of a project's jobsite organization depends upon the culture and style of the corporation, the size of the project, its complexity and contract terms, and the jobsite location with respect to other projects or the contractor's home office. The cost of the project management organization is considered jobsite overhead and must be kept economical to ensure the cost of the contractor's construction operation is competitive with other contractors. The jobsite overhead costs are also referred to as indirect costs or general condition costs. The goal in developing a project management organization is to create the minimum organization needed to manage the project effectively.

General contractors organize their project management teams in either one of two models. In one type of project management concept, estimating, cost control, and scheduling are performed in the contractor's home office by staff specialists, as illustrated in Figure 2.4. In an alternative organizational structure, the estimating, cost control, and scheduling functions are the project manager's responsibilities, as

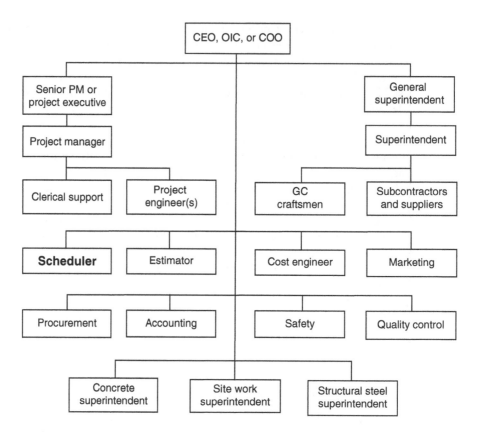

Figure 2.4   Home office staff organization chart

illustrated in Figure 2.5. The latter sole-source style of organization is the one used throughout this book where the PM and superintendent will manage scheduling functions. Both the project manager and the superintendent will report to an individual or individuals within the contractor's upper-management organization. The project manager and the superintendent need to work together as a cohesive team, each with his or her areas of specialization, in order for the project to be successful.

Contractor Team Member Responsibilities

Construction team member responsibilities will vary from company to company and from project to project. The *officer-in-charge* (OIC) is the principal official within the construction company who is responsible for construction operations. He or she generally signs the construction contract and is the individual to whom the project owner turns in the event of any problems with the GC's team. The OIC may also be the vice president for operations, chief operations officer, district manager, senior project

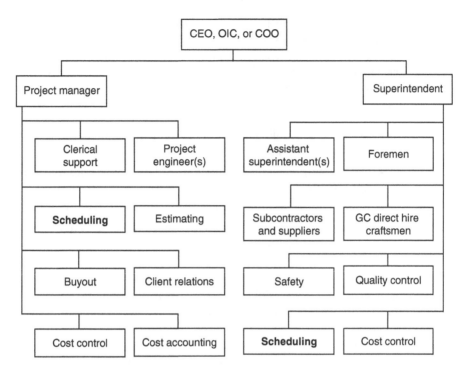

Figure 2.5 Project-based organization chart

manager, or may be the construction company owner or chief executive officer. In the case of a small contractor, these may all be the same person, who may also be the PM and/or superintendent.

The *project manager* reports to the officer-in-charge and has overall responsibility for completing the project in conformance with all contract requirements. He or she organizes and manages the contractor's project team. The focus of this book will be on the functions of the GC's project manager in conjunction with jobsite scheduling. Specific responsibilities of the PM include:

- Coordinating and participating in the development of the project budget and schedule.
- Developing a strategy for executing the project in terms of what work to subcontract.
- Negotiating and finalizing contract change orders with the owner and sub-contractors.
- Submitting monthly progress payment requests to the owner.
- Managing financial close-out activities, and others.

The *superintendent* is responsible for the direct daily supervision of construction field activities on the project, whether the work is performed by the contractor's direct craftsmen or those employed by subcontractors. On larger projects this is delegated to and accomplished by last planners, or those directly responsible for the work, such as assistant superintendents and/or foremen. Specific GC superintendent responsibilities include:

- Planning, scheduling, and coordinating the daily activities of all craftspeople working on the site.
- Determining the building methods and work strategies for construction operations performed by the contractor's own workforce (means and methods).
- Implementing jobsite controls to endeavor to achieve budget and schedule goals.
- Ensuring all work performed conforms to contract requirements.
- Ensuring all construction activities are conducted safely, and many others.

Project engineers or *field engineers* (PEs or FEs) typically report to the project manager and are responsible for coordinating daily details relating to field construction and documentation. On small projects, the project engineer's responsibilities may be performed by the PM. On large projects, there may be multiple PEs. Specific PE responsibilities include processing submittals and requests for information and maintaining associated tracking logs; preparing contract documents and correspondence and maintaining the contract file; and reviewing subcontractor invoices and requests for payment. The PE may also be involved in schedule maintenance.

If the project is of sufficient size or remotely located or of a contract nature that warrants a *jobsite planner* or *scheduler*, that person will have similar responsibilities and background as the project engineer, and may be called a project engineer, but with primarily a schedule focus. The jobsite scheduler will work closely with the project manager to prepare a variety of jobsite and home office planning and reporting tools. The jobsite scheduler will also work closely with the superintendent and help prepare three-week and specialty schedules, among other typical project engineering responsibilities. Smaller to mid-size projects utilize the services of a home-office scheduler. The scheduler visits the jobsite once weekly for the case study project and exercises utilized throughout this book. Just as construction managers and supervisors can have different titles in different organizations, so do planners and schedulers as explained in the next example.

Foremen are "last planners" who report directly to the superintendent and are responsible for the daily direct supervision of craftsmen on the project. The construction firm will assign foremen for work that is performed by the company's own construction craftsmen. Foremen for all subcontracted work will be assigned by each subcontractor. Specific responsibilities include: Coordinating the layout and execution of individual trade work on the project site; verifying that all required tools, equipment, and materials are available before work commences; preparing three-week schedules and sharing them with other GC and subcontractor last planners;

Example 2.1

An international architecture, engineering, and construction firm had many divisions, which were tasked to perform checks and balances on each other. One division included cost and scheduling engineers. These schedulers who scheduled the work of the architects and design engineers were called "planners" but they were not really planning as presented in this book. Rather they scheduled the design portion and their division had one "scheduler" who connected the design effort to construction. The scheduling engineers who were stationed on jobsites were known as schedulers.

and preparing daily or weekly time sheets for their own crews. Last planners are also involved in *pull planning*, which is an element of lean construction and will be discussed later in the book.

This book will attempt to stay gender-neutral and will utilize him or her, he or she, they, them, it, project manager, superintendent, scheduler, and other titles where possible. Some construction terms still remain masculine such as *tradesmen*, *craftsmen*, *foremen*, *man-hours*, and *manpower*, but there are of course many successful women in construction, at all levels in the field and the office. The shortened terms *trades* or *crafts* may also be used where appropriate.

2.8 SUMMARY

There are four major project delivery methods; the primary differences among them are the relationships between the three principal project participants. Owners select contractors by one of two methods, bidding or negotiating. Public owners often are required to openly bid projects and private owners may choose whatever procurement system they are comfortable with. In both cases, contractors may go through a pre-qualification process that shortens the list of firms bidding or proposing. Project owners also determine the method the project is to be priced by the general contractor. The primary pricing methods include lump sum, unit prices for heavy civil projects, and cost-plus fee projects. Many cost-plus projects also have a guaranteed maximum price that financially protects the project owner on the high side but provides cost savings opportunities if the contractor under-runs its estimate.

The discussion of estimating in this chapter was not meant to be a self-contained treatise on estimating, but rather an additional kick-off to some of the backup necessary to prepare construction schedules. The schedulers could not properly do their jobs without a detailed construction estimate. There are many complete and detailed books on estimating, including *Construction Cost Estimating*. The interested reader should look to a resource such as that for a more thorough coverage of this important construction management building block. The construction contract describes the responsibilities of the owner and the contractor and the terms and conditions of their relationship. A thorough understanding of all contractual requirements is essential if a contractor expects to complete the project successfully. Many of these contract considerations as related to scheduling are addressed in Chapter 6.

The contractor's project manager is the leader of the jobsite construction management team. He or she is responsible for managing all the activities required to complete the job on time, within budget, and in conformance with quality requirements specified in the contract. The major phases of a construction project are planning, start-up, control, and close-out. Contractors establish project management organizations to manage construction activities. The GC's project team typically consists of a PM, superintendent, PE, foremen for self-performed work, and administrative support personnel depending upon project size and complexity. Larger or remote projects, especially those with open-book cost and schedule reporting requirements, may also have a scheduler and/or cost engineer located at the jobsite. Each construction project is unique and may utilize a variety of CM formats, processes, and procedures, as introduced in this chapter and reflected in Figure 2.6. The following quote was received from an industry partner:

> *The jobsite team member who understands the schedule*
> *knows more than anyone else.*

Some authors have stated that project management solely involves schedule development and control of the schedule. Although scheduling, as is estimating, is an essential part of project management, there are many other important factors to both as discussed throughout this book. This chapter was not meant to be a comprehensive coverage of CM or PM, but just a brief introduction to a variety of terms and processes that connect to a study of construction planning, scheduling, and control. This concludes introductory Part I and it is hoped the reader is excited to explore the many elements of *planning* in the next part of the book. In addition to schedule planning, there are several CM "plans" developed during the preconstruction phase and process that will be discussed. Lean construction planning has immediate effects on

Figure 2.6   Construction management matrix

estimating and scheduling, as do several elements of the contract that warrant careful consideration by the contractor before submitting a bid or proposal; this is also covered in a later chapter.

2.9   REVIEW QUESTIONS

1. Why would a GC choose to locate their scheduler at (A) the home office, and/or conversely (B) the jobsite office?
2. Why would a client choose one of the following pricing methods? (A) Lump sum, (B) Unit-price, (C) Cost-plus, or (D) GMP
3. What are the differences between an agency CM and an at-risk CM?
4. What criteria would a project owner use to choose a GC on a bid project, and/or conversely a negotiated project?
5. How much more responsibility does a GC PM have in a sole-source organization than in a company that relies heavily on its staff organization?
6. Jobsite overhead costs may also be referred to as ___ costs and/or ___ costs.

2.10   EXERCISES

1. Draw an organization chart for (A) Mountain Construction Company and the case study project, or (B) a project you have been working on outside of the classroom. Include at least 10 companies, positions, and/or individuals. Who performs scheduling? Make whatever assumptions as necessary.
2. In addition to the engineers and consultants listed as built environment participants in this chapter, list one additional firm or type of firm.
3. Subcontractor types were not discussed specifically in this chapter. List three that would be on (A) a typical commercial project, or (B) a single-family house, and/or (C) a heavy civil project.
4. Add one additional responsibility for each of the GC team members described in this chapter.
5. Of the engineers and consultants listed as built environment participants in this chapter, which one (A) is the owner likely to need first, (B) will the architect hire first, and/or (C) would the GC look to when it begins the detailed estimate and schedule development?
6. What five contractors or suppliers might an experienced project owner contract with direct, outside of the control of a typical GC, under a multiple-prime delivery method?
7. What are the advantages for an owner to utilize a design-build delivery method and a cost-plus percentage fee pricing scenario versus a traditional GC delivery method and lump sum bidding? What are the advantages for a contractor for the same scenario?

8. Why would a GC choose to bid a project? There are many potential reasons.

9. When is it acceptable for a project owner to let a project to bid? There are many potential reasons.

10. Has your construction experience been more closely aligned to the organization chart shown in Figure 2.4 or Figure 2.5? As a PM, which organization style would you prefer? As a scheduler, which organization style would you prefer?

Part II

Planning

Preconstruction

3.1 INTRODUCTION

In Part I the reader was introduced to the role of the scheduler, the case study project, and the broad topic of construction management. Now in Part II many planning efforts are elaborated on, including those prepared during the preconstruction (precon) phase, planning the construction schedule, and lean construction planning. In addition, there are several time-related contract issues that must be incorporated before a bid or proposal is forwarded and these considerations are also part of a contractor's preconstruction focus. Construction projects do not just start and end, but rather they have many clear and defined steps or phases. The major phases of a construction project include:

- Planning or preconstruction
- Start-up
- Controls: Cost, schedule, quality, safety, and document control
- Close-out

During the planning phase, the general contractor's (GC's) project manager (PM), superintendent, and scheduler evaluate the risks that are associated with the project, particularly those related to safety, cost, quality, and schedule. Risk analysis and risk management are critical skills essential to successful project management and are elaborated on in Chapter 6, along with other contract considerations that affect the schedule. The PM develops the organizational structure needed to manage the project and the communications strategy to be used within the project management organization and with other project stakeholders. Many material procurement and subcontracting strategies are also developed during the planning phase.

The specific focus in this chapter is on preconstruction. Preconstruction for the project owner includes everything from property purchase and entitlements, to early pro forma estimates and assembling the design and construction teams. In addition to the design team's preparation of design documents during precon, they also are involved in zoning analysis and permit applications. There are essentially unlimited aspects of preconstruction, but this chapter captures those most important for the construction project manager, scheduler, and preconstruction team. The discussion in this chapter includes:

- The preconstruction phase
- Preconstruction services, including estimating, scheduling, constructability review, value engineering, early bid packages, and quality and safety control planning
- Preconstruction contracts
- Preconstruction fees or costs

In addition to scheduling, there are several preconstruction activities that the project manager and superintendent undertake prior to starting the construction of a project. Some of these tasks may be required by the project owner in the contract, while others may be undertaken to ensure a successful project from the contractor's perspective. In the construction manager at-risk project delivery method, the project owner may choose to execute a preconstruction services contract with the contractor. In the design-build project delivery method, preconstruction services are performed as part of the design-build contract. If the project design is completed before the GC is selected, such as in the design-bid-build project delivery method, there is little to no opportunity for the contractor to perform preconstruction services.

The general contractor's role in preconstruction is typically understood to include estimating and scheduling, but there may be many other potential roles. Building information models (BIM) are required by many project owners to provide tools to facilitate collaboration in the design and construction of their projects. Many owners are establishing sustainability goals for their projects that need to be considered in preconstruction planning. Many contractors are also adopting lean construction practices to minimize waste in project execution. One of the techniques being adopted for some projects is off-site construction to prefabricate components that are later installed in the project. Lean construction planning is the focus of Chapter 5.

Scheduling is not the only focus of the construction team during the other phases of a construction project as listed here, but each of these other phases utilize various forms of schedules as discussed throughout this book.

3.2 PRECONSTRUCTION PHASE

So when does the preconstruction phase occur? Obviously it occurs "pre" the construction phase, as shown below:

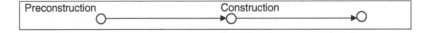

How long does preconstruction last? Well, like the answer to most questions, that depends. Preconstruction for a GC can last as short as a few months or longer than a year. Now the more important question: What is done during the preconstruction phase? That answer is quite long and is client/architect/GC/project specific.

This chapter includes preconstruction activities that are most relevant for the PM, scheduler, and the balance of the jobsite team.

Design Phases

All elements of any given project are not designed at the same time; different elements are started earlier, and some proceed faster than others and include additional levels of detail. There are five major design phases that most built environment projects experience. In some cases, programming and conceptual design may be combined into one phase for a total of four phases. These design phases influence all of the other introductory construction management topics discussed in the last chapter including delivery, procurement, contracting, and pricing methods differently with respect to when the contractor estimates the project and when and how the contractor is chosen and begins their planning and scheduling processes. The typical design phases include:

- Programming,
- Conceptual design,
- Schematic design (SD),
- Design development (DD), and
- Construction documents (CD).

Preconstruction for contractors can occur any time during the preconstruction phase. When (or if) a GC becomes involved in preconstruction depends upon the project owner and the lead designer and the type of project. A simple project, which may be bid lump sum under a traditional delivery method, likely may not have any role for a GC during preconstruction, especially if the client is a government agency. A complicated project such as a hospital may involve the contractor very early during the design process, possibly at the end of SD. When would contractors like to be involved in preconstruction? As early as possible!

Project owners may engage either a general contractor or construction manager (CM) or a preconstruction services contractor or consultant to work during this phase. On some public projects, jurisdiction-dependent, the owner may be required to engage a preconstruction agent who will not end up as the construction contractor. Private negotiated project clients have complete freedom whether to engage a preconstruction agent or not, often with guidance from their architect. Preconstruction services provided by a GC are more prevalent on privately funded projects than they are on public projects. This is because private owners are not restricted to the public bidding procedures that are required of many public owners. For the balance of this chapter's discussion it is assumed that a private client has chosen to engage a GC to perform preconstruction services with full support of the project architect. Also, by participating in precon on a private project, the GC has not necessarily eliminated itself from the possibility of becoming the construction contractor; which is its ultimate goal.

3.3 PRECONSTRUCTION SERVICES

The project owner may choose to select the construction contractor during the development of design and ask the contractor to perform preconstruction services. For contractors this involves attending design coordination and review meetings and providing advice regarding the use of materials, systems, and equipment and cost and schedule implications of design proposals. It is customary to have weekly or twice-monthly preconstruction meetings, often chaired by the architect, which include agendas and action items and meeting notes, very similar to the weekly owner-architect-contractor meeting that occurs throughout construction of most projects, regardless of type. The GC's PM or preconstruction manager may also offer to chair preconstruction meetings. Owners hire GCs or CMs to perform preconstruction services to provide construction expertise often during the DD phase to optimize cost, schedule, and constructability input prior to bidding and during construction. These services may include:

- Preliminary budgeting or estimating;
- Preconstruction phase scheduling;
- Preliminary construction scheduling;
- Constructability analysis, including design document quality control reviews;
- Partnering;
- Building information modeling;
- Sustainability planning;
- Environmental compliance;
- Planning, including development of project-specific quality control (QC) and safety control plans, jobsite layout plan, traffic plan, and others.

Each of these areas is expanded on in this chapter. As indicated in the introduction, there are numerous functions a precon agent or contractor may assist with. Experienced contractors may also participate in a variety of more-advanced preconstruction services such as:

- Management of the design team;
- Material recommendations and selections, including mechanical and electrical systems;
- Early release of subcontractor and supplier bid packages to facilitate long-lead material procurement;
- Permit coordination and expediting;
- Lean processes such as target value design, pull planning, and site logistics planning;

- Value engineering; and
- Other CM considerations such as cash flow predictions and life-cycle cost analysis.

Budget Estimating

One of the main preconstruction contributions the general contractor makes is with early budgeting. The project owner needs to know approximately (within 10–20% accuracy) what the project will cost before they commit financial resources to progress the design to the next phase. Preliminary cost estimates are developed using conceptual cost estimating techniques and refined as the design is completed to ensure the estimated cost of the project is within the owner's budget.

The quantity of budgets and schedules (one, two, three, or more) produced by the contractor's preconstruction team should be defined in the preconstruction contract as discussed later. It is customary for the GC to produce a budget after completion of each of the design phases. Once the DD drawings are finished, the GC will take three to four weeks and develop as detailed an estimate and schedule as is possible. These planning tools will be presented to the project owner and design team with a list of changes from the prior design issue and a list of recommendations to enhance the design either for constructability, cost savings, or schedule enhancement. The owner will then approve and the design team progresses to the next level of design, in this case, CDs.

Often the contractor will maintain a *budget options log* throughout the design process which is very similar to the value engineering (VE) log that will be discussed as a part of lean construction planning. The client and architect may ask the GC to price options or propose changes in the design at each of the weekly design meetings. In this manner everyone is kept as up-to-date as possible with design progression and will not be surprised with large budget swings at the next formal design submission.

The contractor also has an incentive to develop detailed and accurate budgets, even if the design is not yet complete, and make them as true as possible. A construction contractor is not in the sole business of performing preconstruction services; rather, their end goal is to attain a construction contract on this project. The best way for them to do so is to be cooperative team players and produce accurate budgets with minimal swings in value, i.e. no surprises when the final budget, which is often a guaranteed maximum price (GMP), is presented.

Scheduling

Production of construction schedules follows the same course as production of budget estimates. The more detailed the design, the more detailed and accurate will be the schedule. A preliminary construction schedule may be developed to assess the time impacts of design alternatives. In addition to scheduling the procurement and construction activities, the scheduler will also offer to incorporate design and permit activities into their schedule. This is inherently true for design-build construction projects. Some of the types of construction schedules contractors prepare include:

- Precon activities schedule, including design and permits (see Figure 3.1);
- Detailed;
- Summary schedule;
- Three-week look-ahead schedules;
- Subcontractor, area (floor or wing), pull planning, and phased mini-schedules.

Numerous construction contracts, contract special conditions, and technical specifications have been reviewed in preparation of this scheduling book. At various times quotations or paraphrases will be included that relate planning, scheduling, and control to the contract and other various construction management functions. The ConsensusDocs family of contracts was developed by a consortium of built environment associations, including general contractors, construction managers, specialty contractors, suppliers, and other agencies and organizations to replace the previous Associated General Contractors contracts. Contract excerpts from the American Institute of Architects (AIA) contract family will also be referenced in the book. *The ConsensusDocs 500 Standard Agreement and General Conditions between the Owner and Construction Manager where the Construction Manager is at risk* is typically used on negotiated projects and assumes the CM will also perform preconstruction services. That document states in part the following from Article 3.3, Preconstruction Services:

> The construction manager shall prepare a schedule of the work for the design professional's review and owner's approval, it further adds . . . the schedule . . . indicates . . . milestones, material deliveries, shop drawings, phases if appropriate, and other requirements. The construction manager will recommend a schedule for procurement of long-lead items and will help expedite deliveries . . . during the preconstruction period . . .

The final construction schedule will not be prepared until the final set of drawings is issued. Ideally the complete and final cost estimate will have been prepared, which includes direct labor hours for the GC's crew and firm subcontractor prices or quotes. Once the subcontractors have been selected they also will input to the GC's schedule. Planning and development of the formal schedule will be discussed in Chapter 4 and several other sections in this book. The contract schedule is one of many CM products produced from the construction estimate.

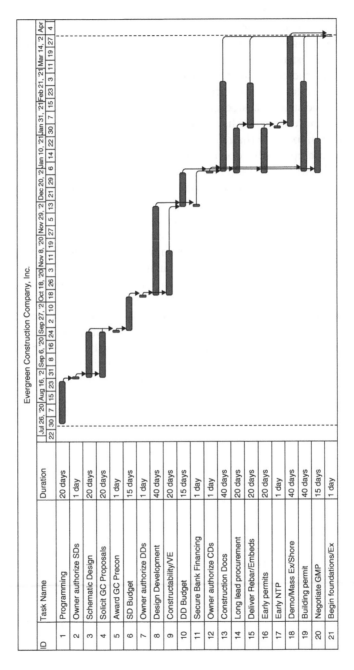

Figure 3.1 Preconstruction schedule

Constructability Review

One important contribution contractors make during the preconstruction phase is to review progress drawings from the design team and comment and make suggestions regarding their "constructability." This is not to say that the design team's documents are not constructible, but rather can they become more constructible or easier to build? Constructability analysis involves reviewing the proposed design for its impact on cost, time, and ease of construction. These proposals are often as simple as changing a welded structural steel connection to bolted, which can be assembled in the field faster and more safely. Or having steel gusset plates welded on to the columns in the fabrication shop rather than in the field, which is also safer and ensures better quality control. An example with wood framing is to change dimensional lumber to engineered lumber, which will be straighter and not shrink. A popular change with wood-framed apartment buildings today is to have the wall and floor systems "panelized" and built in a fabrication shop, which improves quality and enhances the schedule. These prefabricated panels are then flown in with a tower crane and secured with fewer field connections. Many of these types of changes save time and cost, but not all. Some constructability changes may actually increase cost, but improve the schedule, quality, safety, energy efficiency, and/or long-term building maintainability.

The ability for a contractor to have an influence on the design is much greater early in the design process, such as during the SD and early DD phases, but is dramatically reduced during the later CD phase and beyond, as reflected in Figure 3.2.

The construction team also provides another set of eyes and helps edit the drawings. The GC can help mitigate potential subcontractor change order opportunities while reviewing progress drawings. Most designers are not keen on contractors finding errors in the drawings during the construction process and drafting requests for information followed by change order proposals, but they are often appreciative of the GC when errors are corrected before the documents have been let for subcontractor bids.

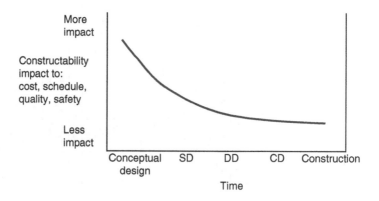

Figure 3.2 Constructability impact curve

Partnering

Successfully executing a construction project involves the participation of many people representing numerous organizations. The ability of the GC to produce a successful project depends on close working relationships among project participants. These relationships generally are defined by contract language, which historically has created adversarial situations. The result of this lack of teamwork often is seen in contract disputes and resulting litigation. Owners, designers, contractors, and subcontractors are beginning to recognize that there is a better way to deliver projects. By working together in a cooperative attitude, many potentially strained relationships can be avoided, and participants can walk away from a completed project with a good feeling, knowing they all have been successful.

Partnering is a cooperative approach to project management that recognizes the importance of all members of the project team, establishes harmonious working relationships among them, and resolves issues in a timely manner, minimizing impact on project execution. The team focuses on common goals and benefits to be achieved collectively during project execution and develops processes to keep all parties working toward these goals. Partnering does not eliminate conflict, but assumes all participants are committed to quality and will act in good faith toward issue resolution. Successful partnering requires an attitude shift among project participants. They must all view the project in terms of making it a collective success, rather than from their narrow parochial perspectives. Mutual trust and commitment are the trademarks of a successful partnering relationship.

Partnering has been used successfully to build cooperative relationships between the owner, designer, and contractor on construction projects. Project results, however, often have been unsatisfactory because major subcontractors were not included in the partnering structure. Subcontractors typically perform the majority of the work on a construction project and have a great impact on project cost, duration, and quality. Major subcontractors must be included in the partnering structure if a cohesive project delivery team is to be forged.

Partnering will work only when there is total commitment from all participants to mutual trust and open, frequent communications. This commitment must originate with the leaders of all organizations represented on the project team. Unless there is commitment and support from the top, mutual trust will not permeate down into each organization represented on the project. This requires clear articulation regarding partnering intentions and a willingness to participate with initial partnering activities and succeeding periodic evaluations. The decision to partner on a project usually rests with the owner and requirements will appear in the special conditions of the contract. Some of the elements of a formal partnering process include:

- A commitment to the partnering strategy is required from top management of each participating organization.
- A workshop is conducted by an external facilitator.

- A team-developed charter defines the project team's mission and goals to achieve project success.
- The project goals should be specific and measurable.
- The communication systems requiring timeframes for responding to inquiries should be established.
- An issue resolution system must be developed that focuses on decision-making at the lowest levels in participating organizations.
- Periodic evaluation of project performance is essential.

An additional service a general contractor brings to the preconstruction team is the ability to solicit pricing from long-lead material subcontractors and suppliers. General contractors rely on positive long-term relations with these companies to receive this valuable early input. These relationships are examples of informal partnering.

Building Information Modeling

Building information models or modeling are useful tools for planning the execution of a construction project through information sharing with designers, subcontractors, and suppliers. BIM allows traditional two-dimensional drawings to be viewed in three dimensions. These documents and processes enable visualization of the project during early phases of construction as well as when it is completed. The models can be used to determine constructability conflicts among the various disciplines involved in the design, such as structural, architectural, and mechanical. The BIM files are compared with each other to identify conflicts. The models also enable visualization of the work to reduce uncertainty, improve safety, resolve scheduling issues, and plan for the use of prefabricated components.

Four-dimensional (4D) models, in which time is the fourth dimension, can be used to simulate construction operations for trade coordination, jobsite safety planning, and creation of pull planning construction schedules. The GC's project superintendent can use the 4D models to simulate the sequence of concrete form construction as well as the placement of concrete. This will enable the superintendent to plan the most efficient sequence of work. The superintendent can also simulate the erection of the basic structure and crane placement to determine the optimal sequence of work. The 4D models are effective communication tools to schedule material deliveries and identify temporary material storage areas on the site. An example of a 4D BIM is included as a figure in Chapter 12.

Another use of the model is to design building components that can be prefabricated remotely and installed on site as assemblies. Building information modeling files contain parametric modeling information, and many fabricators use 3D models to build their components. Use of off-site prefabricated techniques are also discussed in Chapter 5. Project managers need to anticipate the use of BIM technology during preconstruction planning. Such technology can be very useful in planning and scheduling construction activities.

Sustainable Construction

The objective of sustainable construction is to plan and execute a construction project in such a manner as to minimize the adverse impact of the construction process on the environment. While much of the sustainable aspects of a completed project are a result of design decisions, other aspects are due to the manner in which the construction is performed.

Many project owners are interested in constructing sustainable buildings and seek to obtain certification from the U.S. Green Building Council under its Leadership in Energy and Environmental Design (LEED) building assessment system. There are different LEED rating systems for various building types and projects. Certification of LEED is based on a maximum of 110 points distributed across eight categories. The level of certification achieved, from "certified" to "platinum," is a function of the number of points earned by the project. Many cities now require sustainability design and construction methods as part of their local building codes and compliance is a condition of receiving a building permit.

Early in preconstruction planning, the general contractor must consider LEED requirements in the selection of materials, subcontractors, and construction strategies. Material reuse minimizes construction waste as well as earns material and resource credits. Documentation will be needed to demonstrate achievement of certain credits and submission of needed documentation must be included in material supply purchase orders and subcontract agreements.

Environmental Compliance

Depending upon the location of the construction project site, there may be multiple environmental restrictions placed on construction operations. There may be noise restrictions at night or the site may be near a protected body of water or wetlands. In most instances, the general contractor will be required to control soil erosion and stormwater runoff. An early submittal required by the project owner or the city would be a project-specific stormwater pollution prevention plan. This plan identifies potential sources of stormwater pollution on the construction site and identifies measures to be implemented to eliminate polluted stormwater from leaving the site. This often means taking steps to capture the stormwater or retain it to enable infiltration into the soil. The quantity of soil erosion is influenced by the climate, topography, soils, and vegetative cover.

Many projects require building demolition that may involve removal of materials that contain hazardous waste, such as lead-based paint or asbestos. Proper documentation and disposal requirements need to be understood by all parties involved in the removal of the hazardous waste. Spill prevention plans are often required to reduce the potential for contaminating the soil due to construction operations, such as fueling equipment. Many hazardous materials may be used in construction operations, and any excess hazardous materials must be disposed of properly.

Planning

The words "plan" or "planning" have many different connotations in construction. To *plan* is a broad term that means to look ahead or to anticipate what is coming, which is a trait of construction leaders, including jobsite supervisors. As discussed in this chapter, contractors develop several plans or preconstruction plans showing how they anticipate managing the project. A mid-size commercial GC on a negotiated project will typically develop the following plans during preconstruction:

- Quality control plan,
- Safety plan,
- Jobsite layout plan,
- Traffic plan,
- Subcontract buyout plan,
- Permitting plan,
- Start-up and close-out plans,
- Cost control plan, and others.

A well-thought-out plan does not eliminate problems, but it reduces their potential impacts to the project. Planning, essentially, flushes out problems before they occur. Without a plan, construction companies and projects will realize waste, accidents, surprises, required rework, cost increases, time increases, and conflicts within organizations – all at corporate, project, and personal levels.

The contractor's schedule, or bar chart, is thought of as a *plan* that graphically shows all of the construction activities including their anticipated start and completion dates. Built environment participants typically use the term *planning and scheduling* to describe that process. Design drawings are generically referred to as plans; a rolled-up set of drawings will be called a set of plans. City *planning* or urban planning focuses on how best the city should grow and what types and sizes of projects it will allow, including zoning.

Very similar to the constructability review above, early *quality control planning* can have significant impacts on the contractor's ability to meet schedule and cost goals. Contractors can input to early design development documents regarding their ability to meet the client's and architect's intentions. It is important that the QC plan is project-specific and not generic. It should include tasks such as prequalification of subcontractors and suppliers, preconstruction meetings with subcontractors, submittals and mockups, inspectors and inspections, and early in-process punch lists.

Development of an active, not passive, *safety plan* is also a preconstruction activity. Passive safety (and quality) efforts involve fixing problems after accidents occur. A proactive plan prevents issues from happening. Some considerations the active safety plan will address involve full-time safety inspectors, prequalifying subcontractors, Monday morning safety meetings, requirements for personal protective equipment specified in subcontract agreements, bilingual safety instructions, and changing building design to make it safer to build. An example is raising a roof parapet just a few inches such that craftsmen working on the roof both during construction and after would not need to be tied-off or require spotters. Another example is including shop-welded gussets on columns with eye-holes in them allowing cables to be strung serving as tie-offs and guard rail while the structural steel is being erected. Safety plans are often required to be submitted with negotiated proposals or shortly after receipt of a notice to proceed in a competitively bid lump sum project.

The superintendent does not wait until day one of construction to plan his or her site logistics. A *jobsite layout plan* will be developed during preconstruction that will consider material laydown or staging, site access and traffic flow, crane locations, dumpster and trailer locations, stormwater control, and others. The GC's superintendent is the proper person to develop this plan as it is his or her site to manage. There are many contributions the superintendent can make throughout the preconstruction process. An expanded discussion of the role of the project superintendent, both during preconstruction and throughout construction, is included in *Construction Superintendents, Essential Skills for the Next Generation*. The interested reader may wish to look to a resource such as that for additional superintendent roles and responsibilities. Preconstruction plans also include a jobsite organization plan (Chapter 2) and a subcontracting plan (Chapter 17), both of which also involve the superintendent. These and many other preconstruction "plans" developed by the project team have some effect on the anticipated construction timeframe and should be incorporated into early schedules by the GC's scheduler.

3.4 PRECONSTRUCTION CONTRACTS

The adage that fences make good neighbors applies to construction as well. Good contracts make good contractors and good construction projects, and this also holds true with preconstruction. During a slow economy, contractors will offer to perform preconstruction services for free and will not request a preconstruction contract from the project owner. This may be a mistake. The contractor's goal obviously is to get its foot in the door so that they at least have the first shot at a construction contract. A preconstruction services contract is a professional services agreement similar to a design services contract and is not a construction contract. A short preconstruction agreement should be drafted and signed by both parties that clearly defines the expectations from the client and the architect of the contractor and deliverables as discussed above. The cost of these services should be clearly defined and what promises are made, if any, involving a potential construction contract. Some of these precon contract considerations include:

- How many budgets are required, what is the timing, and in what detail?
- How many schedules are required and are they preconstruction phase schedules or construction phase schedules?

- How many meetings is the team to attend, for how long, and who will prepare meeting notes?
- Is travel required to inspect potential material fabrication facilities?
- Are early material submittals expected?
- Are outside workshops such as partnering and lean planned and who pays for these?
- What are the VE expectations?
- What is the anticipated duration of the preconstruction phase?
- Will the GC receive a construction contract at completion of a successful preconstruction phase?

There are copyrighted documents from AIA and ConsensusDocs for preconstruction services but many contractors will offer to draft up a short proposal, defining services and costs and timing, with space for the project owner to sign in agreement. In other cases, the owner will issue a *letter of intent* (LOI) that states the contractor will be reimbursed, for say $70,000, for six months of preconstruction assistance, and it is the project owner's intent to give them a construction contract at the completion of the preconstruction phase. These agreements often state that the contractor will prepare a GMP at completion of preconstruction and if that GMP is acceptable to the owner they will roll their precon costs into a construction contract. But if the parties cannot agree on a GMP, there is no obligation of a construction contract and the contractor will be paid their precon fee and the two firms will part ways. If the GMP is approved, the LOI or preconstruction contract then should be attached to the prime contract as an exhibit, once it is finalized.

Preconstruction services are more prevalent on privately funded projects than they are on public projects. This is because private owners are not restricted to the public bidding procedures that are required of many public owners. Contractors may negotiate a lump sum contract with an owner for preconstruction services or a time and materials contract. The owner must first define the set of preconstruction services desired and then negotiate a service contract with the construction contractor. Some public owners use agency construction management firms to perform preconstruction services. Contractors or construction managers are contracted to perform preconstruction services to provide construction expertise during design development to minimize cost, schedule, and constructability issues prior to bidding and during construction. A sample preconstruction agreement is included on the companion website.

3.5 PRECONSTRUCTION FEES

Preconstruction has a cost. The more the contractor participates in the process, the more deliverables they provide, the longer the preconstruction phase, the more it costs. The amount the contractor charges and the amount the project owner wants to pay also follows economic cycles. During a slow market, contractors will offer to perform preconstruction for free just to get a shot at negotiating a construction contract. During a busy market they may charge $120,000 for eight months of effort, again depending upon the detail expected. Some refer to the preconstruction cost as a fee, similar to designers charging a fee for design services. Contractors prepare an estimate of the preconstruction fee similar to estimating any other work that factors scope, hours, wages, and required materials or resources. The variations of the preconstruction fee options are very similar to estimate types, including:

- Lump sum fee, say $100,000 for 10 months of work;
- Time and material wages with a loaded (including labor burden and home office overhead and materials and profit) hourly fee or rate such as $130 for the PM, $140 for the superintendent, $105 for the scheduler, and so on;
- Hourly fees with estimated hours to come up with a budget for preconstruction services;
- GMP of $75,000 based on hourly rates and hours and quantity of meetings and definition of deliverables expected;
- Preconstruction fee of design-build or design-assist mechanical and electrical subcontractors may be added to the GC's precon fee;
- Description of reimbursable costs either included with or in addition to any of these fee options; and/or
- If long-lead materials are procured, their costs may also be added to the preconstruction fee.

Regardless of the structure of the preconstruction fee, contractors rarely completely cover their cost; they generally do this work at a loss. Again, construction contractors are not in the business of performing preconstruction services; rather they are looking for an opportunity to negotiate a construction contract. But project owners who expect to receive these services for free, and/or without a preconstruction contract, often receive exactly what they pay for and are unhappy with the contractor's contribution toward design completion.

3.6 SUMMARY

All projects realize a preconstruction phase; design is accomplished pre-, or before, construction, but not all projects involve preconstruction services from estimating and scheduling consultants or contractors. Contractors are not solely in the business of performing preconstruction services but often do so in hopes of negotiating a construction contract at the completion of the preconstruction phase. There are a variety of services a GC may perform during this process, including:

- Budgeting or estimating,
- Scheduling,
- Constructability review,
- Long-lead material supplier procurement,
- Quality, safety, and jobsite layout planning, and others.

Successful contractors prepare many preconstruction plans, especially on negotiated projects. But even on competitively bid lump sum projects, contractors must prepare a variety of plans before commencing with foundation excavation. The construction schedule is just one of those documents prepared during a vigorous preconstruction planning effort. Some of the other preconstruction plans developed by GCs include: environmental plan, quality control plan, safety plan, jobsite layout plan, traffic plan, subcontract buyout plan, and many others. All of these plans will later require implementation and control by the jobsite management team, which is the focus of Part IV of this book. Planning requires the contractor to be flexible enough to accommodate unanticipated changes and the result of planning, or the "plan," should include enough flexibility to adjust for changes; there is not a perfect plan. But even though no plan is ever perfect, an imperfect plan – which can be improved upon – is better than no plan at all.

Successful projects require collaboration among many parties representing the project owner, the designer, the general contractor, the subcontractors, the suppliers, and the regulatory agencies. Some project owners include in their contracts provisions for establishing a partnering relationship among these parties. The general contractor, especially the PM and superintendent, must also develop team relationships with specialty contractors.

Preconstruction contracts or letters of intent are good instruments to clearly define the amount the contractor will be paid for its work, the duration of the precon phase, and the deliverables the project owner and architect expect from the construction team. Contractors should be paid for their services, the value of which is rarely sufficient to cover their cost. During a slow economy the precon fee is minimal, as the GC's goal is primarily to get a shot at a construction project. During a busy economy GCs customarily perform preconstruction less often, and when they do, the amount they charge for their services increases.

3.7 REVIEW QUESTIONS

1. Why would a construction contractor perform precon for free? Why would they do so below cost? Would an estimating or scheduling consultant do this work at a loss?
2. Why does estimating typically precede scheduling?
3. What is the difference between an LOI and a precon agreement?

4. As a GC, which would you prefer: a LOI, precon agreement, or construction contract?

5. Match the following estimate and schedule types with the design phases. You can use each answer more than once or not at all, and some design phases will have more than one answer from each category.

Design Phases	Estimate Types	Schedule Types
Programming	(A) GMP	(F) Detailed
Conceptual design	(B) Unit price	(G) Submittal
Schematic design	(C) ROM budget	(H) Contract
Design development	(D) Lump sum	(I) Three-week
Construction documents	(E) Cost plus	(J) Summary

3.8 EXERCISES

1. Why would the GC offer to chair the weekly design coordination meetings and preparation of meeting notes?

2. Draft a preconstruction proposal with a stated scope and fee and duration and include a space for the project owner to sign it.

3. Provide a three-point argument why the owner should hire the GC earlier rather than later during preconstruction.

4. As the architect or project owner, why would you *not* want to hire the GC early during design?

5. Other than the preconstruction services described in this chapter, what service might a consultant or contractor offer the client during design?

6. Assume a major subcontractor, such as mechanical, electrical, elevator, or laboratory casework is added to the precon team. Describe the process when and how they should be chosen, contracted, paid, and what their deliverables would include.

7. Building on Exercise 6, how does the precon process with a subcontractor conclude? How are these situations handled and what are the implications if (A) the subcontractor is dismissed, or, alternatively, (B) they are employed as a member of the construction team?

8. How would your answers to Exercises 6 and 7 change if the subcontractors were (A) design-bid-build, with a separate designer working for the architect, or (B) design-build, where the subcontractors performed their own design?

9. What might cause a project not to proceed as planned, such that the LOI does not roll into a construction contract? There are several possibilities.

10. Looking back to the organization charts presented in the previous chapter, and those which you may have prepared for an outside case study project, list three members from the GC's team who should participate on the preconstruction team and three members from outside of the GC's organization who should be on the team. There are many possible answers for this question.

Chapter 4

Schedule Planning

4.1 INTRODUCTION

The schedule must first be planned before it can be input into the computer and a schedule printed to allow the jobsite team to manage or implement project controls. This chapter focuses on schedule planning. Schedule development is the drafting or creation of the actual schedule document that is discussed in Chapter 8, and then further on the schedule is used as a project management (PM) control tool in Chapter 13. As discussed in Chapter 1, time management is just as important to project success as is cost management. The key to effective time management is to carefully plan the work to be performed, develop a realistic construction schedule, and then manage the performance of the work. Schedules are working documents that need updating as conditions change on the project.

The definition of "planning" is to organize tasks to accomplish a goal. Planning involves placing construction activities together in a logical sequence, which will create a project network. *Webster's Collegiate Dictionary* defines "plan" as the foundation and arrangement of elements, which is certainly a solid connection to construction scheduling. It is important for the scheduler to review the construction contract, drawings, and specifications when developing the plan.

The process of scheduling assigns durations to the plan. The schedule communicates the plan to project stakeholders. A plan without a schedule or a schedule without a plan does not have merit as a tool to assist with building a construction project. Throughout construction the schedule reminds the team that the plan is both the effort that went into developing the schedule, and is a physical document, often a contract. The general contractor's (GC's) project manager and superintendent must be able to communicate the plan with their subcontractors and crew. The ability to communicate is always at the top of the list of traits of construction leaders. The schedule is a valuable communication tool of the jobsite management team. This works two ways – to deliver information and to receive information. The task of the scheduler and jobsite team is to develop the plan, communicate the plan (schedule), and implement the plan (control).

A good analogy for planning is a vacation or road trip. First you decide where and relatively when – that is the plan. Making reservations for the plane, rental car, hotel, and activities to do when arriving, along with getting a map of the area, some background on restaurants, and approval to take the time from work – that is the schedule. Packing, setting the alarm, driving to the airport, parking, taking the plane, checking into the hotel, going out to dinner, taking in the sights – that is the control.

Most scheduling textbooks use examples with activities A, B, and C, whereas this book will use actual construction activities such as form, rebar, and place concrete. Another series of construction activities includes wall framing, mechanical and electrical rough-in, test and inspect, insulation, wallboard, and many others. Use of actual construction activities enhances an individual's ability to learn schedule logic, the means and methods of construction, and the mechanics and calculations of scheduling. The book also utilizes real craft descriptions, such as carpenters and electricians, and real subcontractors, such as earthwork and drywall and others.

This chapter includes discussions of schedule planning elements, work breakdown structure (WBS), logic development, an introduction to resource planning, schedule variables, team collaboration, and concludes with the first rough schedule draft. Additional chapters will build upon this draft and the formal schedule development is discussed in Part III of this book.

4.2 PLANNING ELEMENTS

Planning and *scheduling* are terms many use synonymously, but in this case they represent two distinct phases and products. The schedule is the final product produced by an initial and thorough planning phase. Project planning is the process of selecting the construction means and methods and the sequence of work for a project. Planning must be completed before a schedule is developed. Planning is the hard work and the first step in development of the detailed schedule. It starts with the assembly of all the information necessary to produce a schedule. Some of the steps in planning for development of the project schedule include:

- Developing the work breakdown structure, which is a listing of all the activities that must be performed to complete the project. This should be accomplished before the estimate quantity take-off (QTO) is started. A partial five-level example WBS for the case study project is included in the next section.
- Acquiring input from the key members of the project team including:
  - Field supervisors including superintendents and foremen,
  - Company specialists such as an ironworker superintendent,
  - Key subcontractors, and
  - Suppliers with long lead fabrication times.
- Making decisions with field supervisors regarding:
  - Site layout,
  - Sequence in which work will be performed,
  - Direction of workflow: Bottom up, left to right, east to west,

- Means and methods of construction,
- Type of concrete forming system to be used,
- Material handling, equipment, and hoisting, and
- Safety requirements.
- Making decisions regarding performing work with the contractor's own work crews or with subcontractors.
- Identifying all restraining factors such as:
  - Skilled labor: Crew makeup and sizes,
  - Material and equipment delivery date estimations,
  - Laydown potential,
  - Weather,
  - Permits, and
  - Financing.
- The schedule is not prepared until the estimate has been completed, at least through the work that is to be performed by the contractor's direct workforce. Craft hours from the estimate are needed to determine activity durations for the schedule. The project superintendent should play a critical role in estimate and schedule development.

Figure 4.1 is a schedule planning flowchart, which starts with preconstruction drawing review, WBS, logic, QTO, estimate, computation of direct work hours, superintendent involvement, subcontractor collaborative input, and many rough drafts and edits, among other steps. This flowchart is a continuation of the one introduced in Chapter 1. Additional flowcharts will add more detail in subsequent chapters.

The ability to properly plan for a construction schedule is a unique skill. The entire project is a combination of parts and pieces as reflected in the next example. The project plan may be an outline, notes, minutes of meetings with responsible parties, or a roughed-out schedule. This information is then provided to the scheduler, who in some instances may be the same individual as the planner. On some projects, the PM and/or the superintendent may do the planning and develop the schedule. The preconstruction phase includes estimating and scheduling and other forward-thinking planning processes as discussed throughout this book. Not every PM and superintendent can visualize a completed hotel or oil refinery by leafing through a set of drawings and factor lead times and manpower into their projections. Field managers who have these skills are essential resources for any construction contractor.

The Associated General Contractors (AGC) has a series of books and classes called "Superintendent Training Program" (STP). Unit 3 of that series is titled "Planning and Scheduling." This author has taught the entire STP series of classes to members of the carpenters and electricians unions and contractor associations on numerous occasions. These books are well-suited for the journeyman who wants to

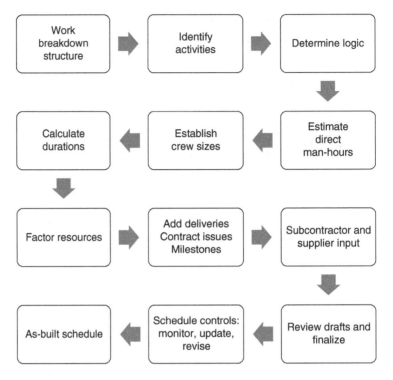

Figure 4.1 Schedule planning flowchart

Example 4.1

This author's father was a residential contractor. He would enjoy monthly trips to his son's large downtown construction projects, but he was in awe of their complexity and size. On each visit he would ask his project manager son: "How does a superintendent know how to do all this?" My response: "He doesn't know how to build it all, he just knows the first step, footings, then he plans the second, the slab-on-grade; planning, scheduling, and building the project one step, or one floor at a time, but he is always looking at the next two or three following steps. The superintendent also relies on his team of experts which includes many foremen and subcontractors and maybe the project engineer or project manager."

become a foreman and for the foreman who wants to become a superintendent. The six major planning and scheduling steps per the AGC, which correspond to much of this book's coverage, include:

1. Identify project activities through the initial work breakdown structure.
2. Estimate activity durations utilizing work days.
3. Develop a project plan that logically arranges the activities.
4. Schedule project activities incorporating the calendar, available resources, and relationships between the activities.
5. Review and analyze the draft schedule, or several draft schedules, and then finalize and formalize.
6. Implement the schedule by communicating to project stakeholders and commence with project reporting and control techniques.

Activities

An activity is defined as a specific task with start and end dates or durations. Webster adds to this definition "action, process, participate, and/or a unit for performing a specific function." Activities require resources and have relations to other activities. Activities are measurable. An activity is a unique definable task with a unique identifier and duration. The choice of planning activities is often derived from the estimator's WBS and follows the 80-20 rule, such that the size of an activity should overly be summarized and not overly detailed. Recall that 80% of the time is included in 20% of the activities. The control phase of a construction project measures activity accomplishments and makes adjustments as necessary.

Activities in a construction schedule are seldom independent occurrences. Most activities cannot start until another activity has been completed, as shown in this footing example. A single activity labeled as "footings" or "foundations" would be acceptable for a summary schedule but is too generic for a detailed schedule. Interaction among construction activities is known as their relationship to one another. Relationships include precedent activities, those that precede or happen prior, and successors or what activities follow another such as:

- First step: Excavate the strip footings,
- Install the formwork,
- Place rebar, including any steel embeds, anchor bolts and seismic hold-downs,
- Place concrete and rod-off,
- Strip forms once acceptably cured, and
- Backfill.

It is recommended the scheduler utilize an activity numbering system which allows new activities to be added to the schedule later and not disrupt the numbering system. For example, many computer software systems default to numbering activities

1, 2, and 3, or A, B, and C. There is no flexibility to insert a new activity in this system. Conversely, if activities are numbered 10, 20, and 30, then a new activity 15 can be added at a later time.

4.3   WORK BREAKDOWN STRUCTURE

The process of planning and developing a schedule was illustrated in flowchart Figure 4.1. The work breakdown structure for the project is an early outline of the significant work items that will have associated cost or schedule considerations. This includes scopes of work such as concrete walls, exterior paving, ceramic tile, electrical, etc. Before detailed QTOs and pricing are prepared, the estimator should develop this general picture of all of the work that will be included on the WBS.

Figure 4.2 is a partial WBS example for the case study project. This list is a good reference to use throughout the estimating and scheduling processes as well as a final checklist to review again prior to finalizing. This early WBS is not to be considered a final product, just a good first step for activity development. The WBS will continue to evolve throughout the estimating, scheduling, and subcontract buyout processes. As the planner dives deeper into the project, there will be several more detailed subsequent levels of the WBS.

The work breakdown structure can be in outline form as shown in Figure 4.2 or with a flowchart/hierarchy. Others utilize an organization chart to represent a WBS. An example of this alternative approach is included on the companion website. The WBS, if properly developed including sufficient detail, can also function as a preliminary "bill-of-materials" from which the superintendent and cost engineer can issue short-form purchase orders and place material orders. The WBS at the detail activity level can be used as cost and schedule control work packages, each with a defined start date, duration, finish date, scope, and responsibility. If the scheduler does not break the project down by activities utilizing the WBS, then there are optional breakdowns available, such as:

- By design discipline, for example: mechanical engineer;
- By drawing type then individual drawing, for example: structural foundation plan;
- By specification (spec) division, then spec section, for example: Construction Specification Institute (CSI) division 05;
- By subcontractor, for example: glazier;
- By craft, for example: electrician;
- Building area, phase, or floor, for example: basement; or
- A mixture of these.

As with many elements of construction management, including estimating, scheduling, and work breakdown structure development, the schedule developer or

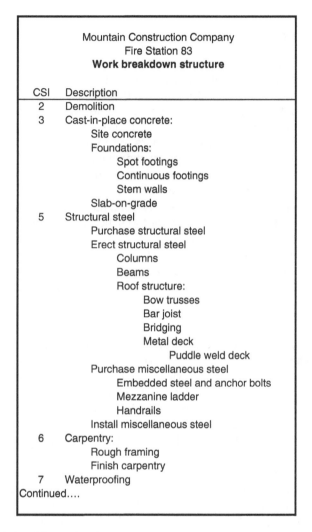

Figure 4.2   Work breakdown structure

planner must remember the 80-20 rule and focus on priorities, which have the most significant impact, as shown in the next example.

Project Item List

Another version of a work breakdown structure is a project item list. The project item list is essentially a WBS with added columns delineating direct labor and material from that of subcontractors. This allows the scheduler (and estimator) to further divide work categories into those self-performed by the GC or those which were subcontracted. The process to develop a project item list is to brainstorm individually, or in

Example 4.2

Every month the mechanical scheduler on this large power plant project prepared a detailed schedule of all the remaining work and presented it to his boss who presented it to his boss and so on, up the chain of command. The critical mechanical work at the time involved rebuilding the pipe hangers in the reactor building. Many of these hanger retrofits would require hundreds of expensive pipefitter hours, often on overtime. But the scheduler was also involved in many other mechanical aspects on this multibillion-dollar project and he would give all the work, from replacing a drinking water fountain to the reactor hangers, equal weight in his schedule. Upon presentation, his boss was not happy with the very detailed schedule. The supervisor was an accomplished artist and drew for the scheduler a "horse in the barnyard" sketch. He started with a beautiful sketch of a horse in an empty barnyard but continued adding water troughs, hay bales, riding equipment, and even a chicken or two. Pretty soon the horse was lost in the clutter, but the lesson learned by this scheduler would not be forgotten 40 years later.

a small group, by leafing through the drawings and noting each work item observed. The specifications are then reviewed and added to this list. There is no exact sequence or order in the initial review. The list can then later be sorted by CSI or spec section or drawing number or GC, and so forth. An example of a project item list for the case study is shown in Figure 4.3 and a blank live version of this form, along with several schedule and estimating forms, is included on the companion website.

Construction tasks that are to be self-performed will require a more intense work breakdown effort. It is appropriate at this early WBS phase for the planner to separate direct versus subcontracted scopes of work. Some of the types of work commercial general contractors may perform with their own direct crews include: Concrete formwork and placement, structural steel erection, rough carpentry, finish carpentry, doors, specialties, and accessories. These are scopes of work often installed directly by the GC's carpenters, laborers, and ironworkers. Decisions regarding which scopes of work to self-perform and which to subcontract are based on several criteria:

- Subcontractors will be used if the specialized tradesmen needed are not employed by the general contractor.
- The reason subcontractors are also known as specialty contractors is that they specialize in a specific scope of work, such as windows, and should be expected to know that work better than would other contractors.
- If there are problems with the quality of installation, a subcontractor is required to repair the work without increase in cost to the general contractor.

			Provider			
			GC		Sub	
Line	CSI Div	Cost item description	Matl	Lab	Matl	Lab
1	w/3	Structural excavation			X	X
2	3	Form, place, and finish concrete	X	X		
3	3	Reinforcing steel supply	X			
4	3	Reinforcing steel installation				X
5	5	Structural and misc. steel fab	X	X		
52	10	Specialties	X	X		
74	23	HVAC			X	X
80	26	Electrical			X	X
88	32	Site utilities			X	X
		Continued…				

Mountain Construction Company
Project item list

Project: Fire Station 83   Date: 5/19/22
Estimator: Charles Kent

Figure 4.3   Project item list

- The subcontractor may have estimated a fixed price that is less than what the GC has estimated; therefore the subcontractor would be absorbing the pricing risk.
- Subcontractors may have craftsmen and equipment available that the GC does not.
- There are risk mitigation arguments on both sides as to which – the subcontractor or the GC – can perform the work faster, safer, cheaper, and with better quality than the other.

4.4 LOGIC

Once the activity list has been completed, it is time to begin development of a construction schedule, or in this case, a logic diagram. Individual tasks or activities to be accomplished were identified during the development of the WBS. The next step is to determine the sequence in which the activities are to be completed. This involves answering the following questions for each activity:

- What deliverables, permits, submittals, and contracts need to be in place before starting each activity?
- What activities must be completed before another activity can start?

- What activities can be started once another activity has started or been completed?
- What activities can be performed concurrently?

Based on the answers to these questions, the schedule structure can be developed using the logic among the activities. Developing the schedule logic is an important scheduling planning step. Logic can be easily described as "What order of activities makes the most sense?" For example, carpenters form the footings before ironworkers place the rebar which is before the laborers place the concrete inside the forms. That is logical. The schedule logic factors dependencies, i.e. what activities are dependent upon one another; this includes predecessors – activities that go first – and successors, activities that follow. The dictionary adds "reason, validity, normative, and predictable" to the description of "logic," and these terms also apply to construction schedule planning.

Even with all the computer sophistication available today, many contractors still develop their logic by hand on butcher paper, or by rearranging colored sticky notes on the whiteboard. The scheduler who focuses on presentation first, and logic later, often misses important steps. Another logic arrangement method is through the use of physical index cards. In addition to the activity description, there are many other entries that can be added to the card, or variables, as discussed below. After the scheduler has arranged all the activity cards in a logical order, he or she must ask themselves, what is missing for this construction project to happen efficiently as planned? Are additional activities needed? Are there missing restraints or significant material deliveries? Missing an activity is a common and critical schedule planning error, more so than finishing one day late on concrete formwork construction. An example of this planning tool is included as a figure later in this chapter.

Relationships

Construction activities do not just happen independently; most, if not all, of them are related to other activities. Essentially, one activity cannot start or finish until another activity happens. These relationships are also known as restraints or dependencies. The roofer cannot install the roofing material until the roof structure has been safely completed. The roof structure restrains the roofing. Roofing is dependent on the roof structure. These are all logical relationships. Restraints are defined as activities that force or have influence or control over another. Restraints prevent or limit other activities.

A *predecessor* to an activity is another activity or milestone that must happen, which includes either starting or completing or both, before another activity or milestone can start or finish or both. An example is simply the fact that the backhoe cannot backfill the footing before the concrete in the footing was placed and forms stripped, among other important activities. A *successor* to an activity is the opposite of a predecessor. A successor activity or milestone cannot start or finish or both before another activity or milestone has started or finished or both. Another example is bolts of the steel beam-to-column connection that cannot be torqued until the column has been erected, among other important activities.

Linking activities, like the links in a chain, arranges events in a logical order based on their relationships. Every activity, except for one, should have at least one restraint or predecessor and every activity, except for one, should have at least one following activity or successor. After the first draft of the schedule (discussed later), the scheduler should check for those activities that are missing predecessors and successors in the logic. "Hanging" or "dangling" activities are those activities plotted without both predecessors and successors. The scheduler should not include any hanging activities other than the very first and the very last activity on the network. If activities do not have both predecessors and/or successors, it begs the question – why could they not have started earlier or finished later? The answer is likely that another relationship or activity such as a material delivery or issuance of permit or inspection needs to be added. There should be only one activity in a complete schedule without a predecessor (the first) and only one without a successor (the last). If that cannot be arranged, a dummy activity may be added, or a milestone with no duration, such as notice to proceed or turnover. The use of dummy activities will be discussed with schedule calculations in Chapter 9.

Lag

Most schedules are conveniently drawn up in a line or series, as the first example in Figure 4.4 represents. But this is because that is the easiest graphical approach. These activities are known to have finish-to-start (FS) relationships because formwork is finished before rebar can start. The FS relationship is typically the network's basic path. In fact, very few activities 100% stop and start in series; most have some overlap as the second example of that figure shows. Overlapping activities may also be known as fast-track construction.

Schedule software allows the scheduler to incorporate overlapping activities with the introduction of lag. Formwork and rebar can then have a start to start (SS) relationship with two days of lag. This essentially indicates the superintendent will start his formwork crew in the southwest corner of the building, following the backhoe, move clockwise north along gridline A, turning east along grids two and three, then south on gridline L, and finishing with the footing formwork heading west on grids seven and eight. The ironworker crew will follow the carpenters two days later in the same path and subsequent crews and activities will follow with similar SS relationships. This is known as the "parade of trades."

A finish-to-finish (FF) relationship reflects whether two activities must finish at the same time, or one finishes slightly ahead of the other, again with a built-in lag. The software can depict all of this but oftentimes the restraint lines become confusing or overlap, and some will unfortunately flow in reverse.

Another approach is for the scheduler to retain the original finish to start relationships but break each activity down into smaller segments, thus eliminating the lag. For example, instead of the schedule showing the carpenter crew forming all of exterior footings as one activity, it would be broken down into:

1. Form gridline A exterior footings,
2. Form gridlines two and three exterior footings,
3. Form gridline L exterior footings, and
4. Form gridlines seven and eight exterior footings.

The other exterior foundation elements, including rebar placement, embeds, placing, concrete, stripping forms, installing perforated pipe drain systems, waterproofing, rigid insulation if appropriate, and backfill would all also be broken down into similar smaller segments. This would then eliminate the need for SS and FF relationships and lag, but it does add considerably more activities to the schedule. This level of detail may be best reserved for very large projects or for three-week short-interval schedules.

Construction activities rarely happen completely in series as reflected in the foundation example earlier. Rather many activities can occur completely or partially in parallel with other activities and have staggered starts or overlapping starts. For some this is known as fast-track scheduling. In schedule terminology this is referred to as "lag." The Fire Station 83 case study project has 520 linear feet of exterior continuous footings. Rather than waiting for each of the earlier steps described above to be 100% completed, they can overlap. A comparison of series versus fast track is shown in Figure 4.4.

Some schedulers also use the term *lag* to reflect a short time period after one activity finishes but before another starts; e.g. "two days of lag." This in a sense is an FS plus two-day lag relationship. Scheduling software such as Microsoft Project easily accommodates this. Some may incorrectly refer to this this two-day lag as *float*. Other scheduling authors and experts may refer to lag as lead-time or delay. But what happens during this lag time (certainly not nothing)? This relationship may be better explained with the introduction of additional activities such as permits, delivery, inspections, concrete cure, testing, material ordering/fabrication/delivery, and others. For example, there's a reason to place concrete on a Friday. The natural two-day weekend lag accommodates curing.

Typically it is easiest to utilize finish-to-start relationships that have zero lag. If the early finish of activity X equals the early start of activity Y, there is zero lag between the two. The use of SS and FF relationships are not as common as some indicate. Most builders will eliminate the lag concept through a variety of options. Activities with SS or FF relationships likely have other smaller activities which they rely on but may be too small to chart and outside of the 80-20 rule. Start-to-finish relations are unusual in construction schedules.

Most contractors will represent a time lag by the introduction of these types of additional detailed activities. Without these additional activities the scheduler may miss some important restraining activities. The introduction of additional or smaller activities is also a method to remove or hide float, which are discussed in Chapter 9.

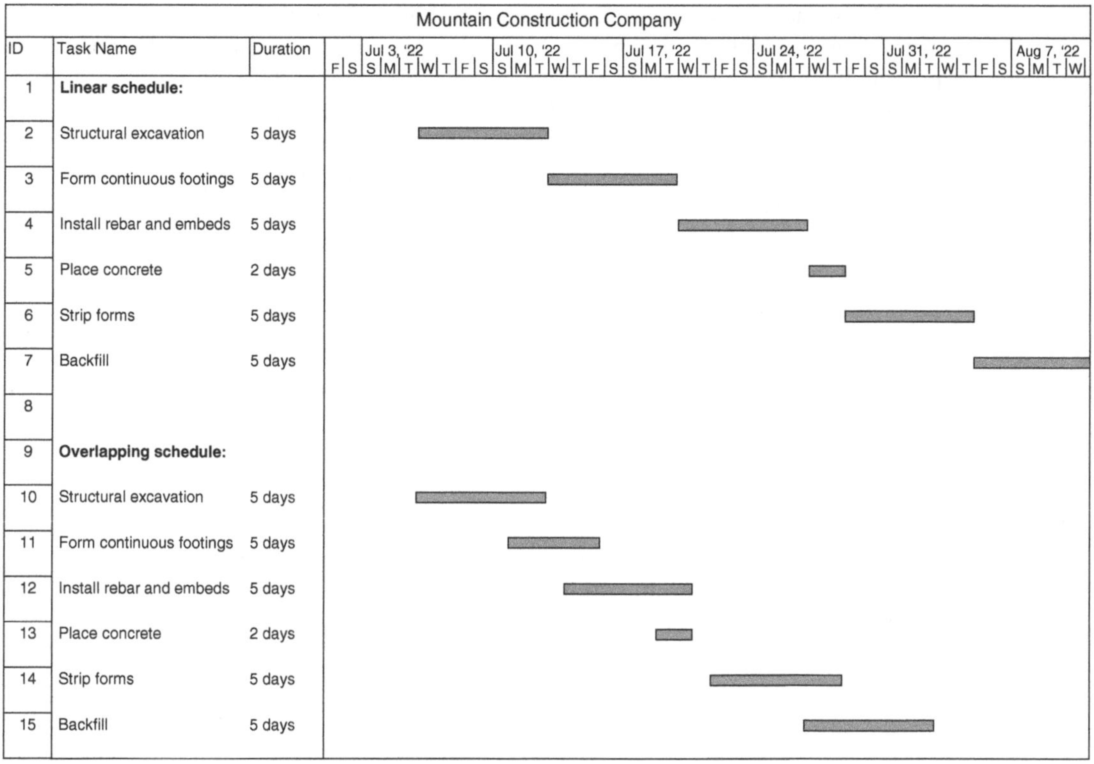

Figure 4.4 Activities in series versus fast-track

Administrative Restraints

There are many reasons why a schedule may not eventually be attained 100% in the field. One of these is the lack of adequate inclusion of restraint activities or milestones. In addition to construction restraints, such as concrete formwork that occurs before rebar, the scheduler should also make note of activities that were not necessarily included in the work breakdown structure or have direct work man-hours associated with them. Where possible, the scheduler should include nonconstruction activities in the schedule logic, especially restraints and those out of the contractor's control. Nonconstruction activities would include administrative activities such as letting a subcontract agreement, processing a submittal, and delivery activities such as fabrication and trucking to the jobsite. The scheduler should think about what must happen before each activity can start or complete and reflect it in the project plan.

Every construction activity has several activities within it. Some of the other non-jobsite construction activities that can be included on a detailed schedule are listed here. Other activities that may be the responsibility of the project owner before a GC can commence include design, securing financing, surveys, geotechnical reports, and potentially demolition and hazardous material abatement. They all must be accomplished before day one on the schedule occurs and should either be noted as an activity with a duration, or a milestone. When a schedule is not met, the contractor is often blamed. But sometimes these delays are out of the contractor's control, or at least out of the field superintendent's control. Built environment managers, including owners and designers, need to know the effect of today's actions on downstream construction activities. Administrative construction activities also include:

- Contracts,
- Permits,
- Shop drawing and submittal creation and approval,
- Material deliveries preceded by fabrication and transportation,
- Inspections, and others.

If elevator work is included as only one activity in a schedule, starting from initial bids and concluding with buyoff from the inspector, it can take as long as a year. The introduction of administrative activities as reflected in Figure 4.5 makes the work easier to plan, schedule, and control.

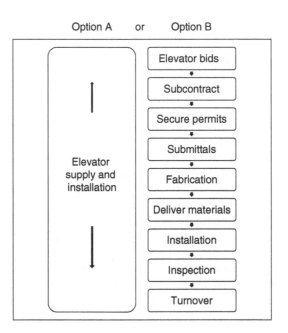

Figure 4.5 Elevator scheduling options

4.5 RESOURCES

Contractors rely on resources to build a construction project. Resources are necessary tools for construction. The lack of sufficient resources would restrain or limit a contractor's ability to construct the project logically and efficiently, as discussed above. Major construction resources include access to the following, which are also the primary areas of an estimate, as discussed prior. Other restraining factors include site logistics, adequate site access, and work hours and days available as dictated by the city.

- Direct labor,
- Construction materials,
- Construction equipment, and
- Subcontractors and suppliers.

The GC and subcontractors both employ "direct" labor in that the labor is directly working on the jobsite and contributing to the construction work, compared to indirect labor, which involves management and supervision. Craftsmen or tradesmen all have unique traits and responsibilities. They are all skilled labor and are not general laborers or workers. A crew or crew members are a group of craftsmen usually working for one foreman. These are the true "builders" and all contribute to building the project. The term *direct labor* as used here refers to the craftsmen working directly for the general contractor. The same concepts also apply to craftsmen working for subcontractors when they are developing their own schedules.

Direct labor is typically considered as the major resource necessary to incorporate into an early schedule. If a contractor does not have access to a reliable workforce, much of the other schedule planning effort is without value. The man-hours a contractor has available are first determined from the detailed estimate as reflected in the planning flowchart in Figure 4.1.

The question schedule planners face is whether to plan the schedule due to availability of resources or to plan the work, develop the logic, produce the schedule, and then load resources to meet the demands of the schedule? For example, in Figure 4.6 the site size and hours available to work, combined with the contractor's workforce, resulted in an available and workable crew size of 30 craftsmen, which is approximately 5,000 hours per month. But according to the schedule, approximately 7,000 hours should be spent in July and August. It is recommended that the scheduler should schedule the project first under normal or ideal circumstances, working standard five eight-hour days with no restrictions on resources and then adjust the schedule or resources to make the two match. Activity durations will be more fully established with schedule development in Chapter 8 and resource balancing will be covered more extensively in Chapter 10.

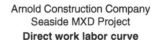

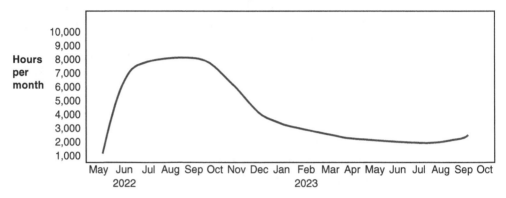

Figure 4.6 Labor requirements curve

4.6 VARIABLES

Does a general contractor utilize an earthwork subcontractor to excavate a site to the top of the footings and later excavate the footings themselves with a backhoe or do they bring the whole site down to the bottom of the footings and backfill up from there? Are footings backfilled only to have the plumbing and electrical subcontractors dig it all up again for under-slab rough-in? Are structural elements prefabricated off-site or assembled onsite if there are available laydown areas? The approach to developing a construction schedule plan is similar to military generals looking over a terrain for what may potentially be a battlefield. How do they want to build this project, or before that, can they build it, or should they build it?

There are many issues the scheduler needs to consider when planning a project schedule. All of these are essentially variables in a very complicated and high-risk mathematical formula – in this case a construction schedule. Some of these considerations include:

- Are the drawings and specifications complete? What do they tell the estimator and scheduler about how they should approach the project?
- Are there *contract* and time considerations that must be incorporated into schedule planning, such as early-occupancy or dual occupancy requirements?
- How many *direct hours* of work are included in the cost estimate? Does the GC have adequate crew sizes developed for both its own and subcontracted work?
- The *jobsite* needs to be visited, hopefully before the project was estimated. What restraints are on or near the site that will impact the schedule plan, such as as wetlands or power poles?

- Are structures required to be *abated* and *demolished* before construction starts, and is that work in the GC's or the owner's scope? Have special permits been applied for? Have specialty contractors and inspectors been engaged? Has a safe off-site disposal site been located?

- Which *superintendent* has been assigned to the project and has he or she also selected their foremen? What specializations do those team members have?

- Have *subcontractors* been selected and have they provided material delivery restraint dates, constraints, manpower estimates, and durations to the GC when developing their estimate and schedule or during a buyout meeting?

- What are the contractor's *means and methods* with respect to workflow, prefabrication, shoring choices, scaffold, formwork, and so on?

- What *equipment* was estimated and/or what is needed? Does the contractor have equipment in-house or do they need to obtain equipment from third-party rental companies? When will equipment be available? Tower cranes often have a lead time.

- Did the city request a *traffic plan* from the project owner as part of an entitlement study? Did the GC submit one with the permit application? Is a preconstruction meeting planned with the city which will impact traffic flow and work hours?

- Is there room on the jobsite for laydown and prefabrication? Did the GC's superintendent develop a *jobsite layout plan* during the estimate process?

- Like traffic, did the city also require an *environmental plan*? Does stormwater run-off need to be protected before mobilization? Is a permanent stormwater detention system (sometimes an underground vault) required to be built before work is started on the building?

- Considering day one of the schedule by contract is June 9, 2022, and the completion date is May 26, 2023, what work will be underway during inclement weather conditions? Should the planner consult a local weather service so that he or she knows how many *rain or snow days* should be incorporated?

- In addition to demolition permits, traffic control, and environmental control rain or snow days the city will have many other conditions the scheduler must incorporate into the plan such as *preconstruction meetings* for shoring. The formal approved permit drawings will have some of these issues noted. The GC's superintendent needs to have those in hand, before a final plan and schedule can be developed. Other special conditions are included in the front end of the spec book and in the contract and the planner must be made aware of any that might have an impact on construction time.

- There are many *nonconstruction activities*, some of them are often critical, which the scheduler should incorporate. These include: subcontract and supplier buyout, submittal preparation and processing, material fabrication and delivery, and others. Many of these fit into the restraint category discussed prior.

Interruptions

Interruptible activities are ones which are not continuous and work must be delayed or halted for a specified time period. The schedule planner can work around these interruptions with the concept of lag. The solution is either the interrupted work is not critical or another activity should be inserted in the middle of the delay. Some potential reasons for interruptible activities include the following:

- A nonworking activity or owner activity, which is common on civil projects,
- Working in occupied buildings,
- A project adjacent to salmon spawning streams,
- Holidays, evenings, or weekends,
- Concrete curing time,
- Inspections, and others.

When activities stop and then restart within themselves, essentially nonworking time, it is likely that something else is happening there, even if not a construction activity and the network may benefit by adding another activity during that time. In essence the activity is interrupted. For example, concrete may require four days of cure time before the forms may be stripped.

Weather is a common source of schedule interruptions. How should a scheduler build in rain days? What if it doesn't rain, is this then float? Should potential impacts be shown in schedule logic? If not, is the contractor negligent? It rains in Seattle, snows in Alaska, tornadoes hit the Midwest, and hurricanes devastate the southeast. A scheduler can build rain days into the schedule by declaring them up front and including them as early activities, showing them as late activities, or adding a day or two here or there throughout the schedule and not calling them out separately. Weather days shown on the schedule can also cause difficulty with schedule updates and schedule revisions. Other related scheduling strategies include hidden logic, hidden float or fluff, and/or contingency time.

Some schedulers use complicated formulas with standard deviations to predict weather impacts. There are too many uncertainties in construction to be this exact; "stuff happens." Imagine you as the project manager explaining the standard deviations you calculated for your schedule to your field superintendent? Reason and logic should always prevail if the schedule is to remain a viable construction tool.

Logic Tools

One planning method is to write activity descriptions on *index cards* (or stickie notes) and arrange them in a logical order as shown in Figure 4.7. This same index card method can be utilized when planning for an "activity on the node" schedule. There are many activity variables that can be added to the index card, beyond the activity description, including activity number or code, duration, predecessors and/or

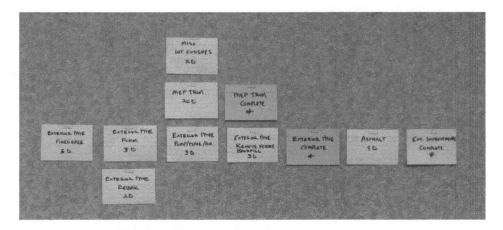

Figure 4.7 Activities on index cards

successors, resource requirements, and both early and late start and completion dates and resulting float calculations. All of these options will be discussed in subsequent chapters.

Assigning *activity codes* or *numbers* in addition to descriptions allows planning to be done easily on a spreadsheet, such as with Figure 4.8. Cost codes and schedule activity codes are necessary but the jobsite management team should not make them overly burdensome, such as the 10-digit codes recommended by some schedulers. The pipe-hanger example discussed in Chapter 15 with controls is what happens when cost and scheduling engineers make it burdensome on the craftsmen to report work progress.

The scheduler may wish to incorporate *milestones* or *constraints* into the logic at this point. Other schedulers may add these dates after the timeline has been incorporated into the schedule and the critical path established, as discussed in Chapters 8 and 9, respectively.

Webster's loosely defines "constraint" as something that is forced or imposed on – in this case, imposed on the contractor. Milestones may also be established by the contractor as a goal. Different authors, professors, and scheduling consultants may mix "constraints" with "restraints" and all likely feel they are most correct. Examples of constraints or milestones that should be noted in the schedule in the context of this book are listed here.

- Contract execution and issuance of notice to proceed,
- Receipt of building permit,
- Mobilization,
- Topping out of structure,
- Dry-in,
- Energize switchgear,

Mountain Construction Company
Fire Station 83

Activity codes

Activity codes to match specification section	Activity description	Simpler activity number
031200	SOG finegrade	10
071255	Vapor barrier under SOG	15
322240	Underslab utilities	20
031000.1.4	Formwork	25
032000	Reinforcement steel	30
033000.2.1	Place/Pump/Finish 8" SOG	35
033000.2.2	Place/Pump/Finish 4" SOG	40
033519	Spray curing compound	45
031000.1.5	Remove forms	50
312000	Backfill at edge	55
033519	Polish SOG	60

Figure 4.8 Activity codes

- Temporary certificate of occupancy and other completion milestones such as certificate of substantial completion and certificate of occupancy,
- Demobilization,
- Close-out, contractual completion, and others.

Schedule planning has many similarities to cost planning, or estimating, many of which have already been discussed. If the contractor establishes several design

Example 4.3

Many of the old-school superintendents did not trust any documentation that came from the home office, and this ironworker turned GC project superintendent was no exception. He would patiently sit with the home office scheduler (the same one as in Example 1.1) and nod in agreement during the day-long schedule development meeting. But the new rolled-up schedule would eventually be left behind the seat in his pickup truck and he would arm-wave project direction as he had always done previously, with admittedly some success, but not always!

or owner activities in its schedule, as milestones or constraints, they are also giving the owner notice that if a permit is not received by a specific date, the schedule will be impacted. Another approach is to provide a list of schedule assumptions that accompany a bid or proposal; for example, the contractor has assumed a typical winter and has incorporated five rain days into its schedule. These qualifications are typically not allowed on public bid projects.

4.7 COLLABORATION

The involvement of the jobsite team in schedule planning and control is emphasized throughout this book. If the PM and superintendent participated in development of the schedule, they will have buy-in and will work hard to see that the schedule is achieved. Conversely field supervisors who are handed a schedule from a staff scheduler or consultant, without providing input, will always have an "out" as to why the project finished late. This philosophy applies not only to the GC's project superintendent, but subcontractor superintendents as well need to be involved in scheduling their own work.

It is a mistake for schedulers to simply plot a schedule without first developing a detailed plan. Technical schedulers who are well-versed in inputting and manipulating schedule activities with scheduling software, but who are not also builders, may be able to print out a realistic-looking schedule, but without a good plan behind it, the schedule is not achievable. The scheduler described in Chapter 1, Example 1.1, was also a builder, and the plans he produced were typically accomplished because he first achieved field superintendent buy-in.

For the plan to be workable, it requires careful consideration of all the activities included in this chapter and buy-in from *all* of the construction team members, including the list below. Example 4.3 reflects what happens without superintendent buy-in.

- General contractor's officer-in-charge,
- Project manager,
- Superintendent,
- Field foreman,
- Project engineer,
- Home office staff support specialists, such as the scheduler, estimator, and quality and safety officers, and
- Subcontractors.

A schedule without the involvement of superintendents and subcontractors is of little value, often reinforced by the saying "garbage in, garbage out." During the bid cycle it is difficult to get subcontractors involved, but like the estimator asking them for an early prebid approximate order of magnitude estimate, the scheduler can ask for an approximate duration. Negotiated projects typically benefit more from early sub-contractor schedule involvement than do lump sum bids. Post award, subcontractors should provide input before the detailed schedule is finalized. Selecting best-value subcontractors involves much more than just low bid. Schedule participation will be discussed in Chapter 17.

4.8 FIRST DRAFT

Once all the logic and resources and other variables have been established, the scheduler still has some work to do. Some will jump right into applying a timeline to their logic diagram at this point and quickly send it to his or her boss or even the project owner. It is best to prepare many rough drafts and mold the plan into a schedule that can later be managed and attained. Drawing a schedule from scratch is not an easy task, especially without going through all the planning steps. It is easier for anyone to mark up others' work with a red pen than it is to take a blank sheet of paper and draft it from scratch; this is a similar analogy to preparing an estimate or writing a book. Using many drafts, the scheduler can more easily secure buy-in from his or her superintendent and their subcontractors. The early bar chart, as discussed in this section and elsewhere, is a common tool to prepare and revise these early draft schedules.

One strategy is for the home office staff scheduler to print the schedule logic, with all activities having the same duration, such as five working days, and then sit around the table with the project team and mark it up, changing from five to two days or five to 10 days. Then the scheduler puts in the new durations and prints out another rough draft, and the team does it again, also adding additional restraints such as material deliveries. It is important to get the logic down first and the durations second and not back into logic just to meet a preimposed milestone date or goal.

Bar Chart

A bar chart is one of the most common schedule formats. The planner often utilizes a simple bar chart to arrange the activities derived from the WBS in a logical fashion. This allows the incorporation of relationships between activities, resource limitations, milestones, or constraint dates placed on the contractor by the project owner or city, and others. As stated, it is important that the schedule be arranged properly, or planned properly, before it is formalized into a contract schedule document.

Bar chart schedules are typically sorted by early start dates. It is customary to format a bar chart schedule in a 'waterfall' effect with the project schedule starting in the upper left-hand corner and completing in the lower right-hand corner of the drawing. A bar chart typically utilizes horizontal lines to represent activities. A bar chart may or may not incorporate vertical "relationship" or restraint lines. Some schedulers cite the lack of dependencies or restraints as a shortcoming of the bar chart format for schedules. But these restraints can be added in and the bar chart in its simplest form is a viable tool for schedule planning and logic development.

If the bar chart arrangement is left solely to the computer, the result may be cluttered and confusing to the users, especially those not accustomed to reading construction schedules. A draft should be plotted, and the scheduler should edit before distribution. It is important to minimize the crossing of vertical restraint lines with activities as much as possible. This may require some slight rearrangement of activity order. Elimination of redundant restraints is also suggested to make the bar chart a good communication tool. Otherwise at times it can appear that there are more vertical than horizontal lines. An example of this is included in a later chapter. A bar chart, which includes vertical restraint lines, can be formalized further by adding a timeline to the logic diagram.

4.9 SUMMARY

There are many elements to schedule planning including identification of construction and nonconstruction activities and their relationships between each other. The schedule planner should "shake the tree" and solicit input from all stakeholders, including the estimator, superintendent, and key subcontractors. The WBS is a valuable tool to develop a construction cost estimate and schedule. It is typically organized by CSI but also should follow the general path of how work will be performed. A project item list expands the WBS by adding columns that delineate direct-installed work from subcontractor-installed. Direct work is the most risky for a contractor and is planned in additional detail.

Logic basically means what work happens in what natural order. The relationship between activities is also known as restraints and involves an understanding of predecessor and successor activities. Construction activities are typically arranged in a schedule based upon one activity finishing before another starts. This is known as the finish-to-start relationship. But seldom are all activities 100% complete before succeeding activities start; they overlap or have lag. Overlapping activities is also known as fast-track. A schedule can minimize this graphically by the introduction of smaller activities, including nonconstruction administrative activities such as submittals or deliveries.

Resources are typically limited for contractors. The availability of labor and equipment resources may affect schedule planning. There are many other variables a schedule planner should consider when developing the schedule logic. If activity interruptions are known ahead of time, they should also be incorporated into the plan. Milestones and constraint dates should also be planned for. A schedule that is developed in a collaborative process will stand a much better chance of achieving success than one developed in a top-down fashion. Collaboration of all key team members including the superintendent and subcontractors is paramount to schedule success.

After the scheduler has incorporated all of these early planning elements, he or she will develop a first rough draft, with or without a timeline or durations. The bar chart is a useful tool for this early draft. It is important for the scheduler to incorporate all activities in a logical order before attempting to squeeze the plan into a predetermined schedule window.

4.10 REVIEW QUESTIONS

1. If a contractor has a staff scheduler or outsources that role, how might a project superintendent provide valuable early planning input?

2. The text introduction to Figure 4.1 listed a few additional key planning efforts. Place them into the figure where they would occur.

3. Why is an activity such as "mechanical work" unacceptable in a detailed schedule?

4. Can you tell from Figure 4.2 which concrete scopes the GC intends to self-perform? Utilizing Figure 4.3, can you now tell?

5. Why might an estimator or a scheduler arrange a WBS or project item list in a logical order?

6. Construction activity relationships are also known as ___ and ___.

7. Why should all but two activities in a network have both predecessors and successors?

8. Does "lag" between two activities represent "float"?

9. What is the difference between a WBS and a project item list?

10. Other than direct labor, what is a typical potential major construction resource restraint that the planner must consider?

4.11 EXERCISES

1. What are some of the immediate (within two to three weeks) downstream activities of (A) drying in the roof and/or (B) installing gypsum wallboard?

2. Expand the outline WBS included in this chapter for another system from the case study project, such as slab-on-grade, wood framing, and/or doors, frames, and hardware.

3. Storefront and kitchen appliance installations are often shown as one activity in a logic diagram but require many predecessors and affect many successors. Choose one system, storefront, or kitchen appliances, or both, and define three critical predecessors and three critical successors.

4. What activities are missing from the string of six strip footing activities discussed in this chapter?

5. What activities are missing from the string of elevator activities included in Figure 4.5? Potentials include both pre- and postconstruction administration activities.

6. Organize these electrical scheduling activities in logical order: (A) Switchgear, (B) Energize, (C) Under-slab rough-in, (D) Wall and ceiling rough-in, (E) Pull wire, (F) Terminate, and (G) Set boxes. If anything is logically missing, feel free to improvise.

7. Place these schedule activities in logical order: (A) Print the schedule, (B) Prepare a WBS, (C) Detailed estimate, (D) Plan the schedule, (E) Status the schedule, and (F) Make a formal schedule revision.

8. Why might an actual labor cost curve have a slight up-tick at the completion of a project?

9. Have you taken any of the AGC's STP courses? Do so if you get an opportunity.

10. Explain how an inadequate, incorrect, or negative answer to each of the variable questions listed in this chapter could be detrimental to schedule planning.

11. Assume you are a superintendent. How would you feel and/or respond to being handed a construction schedule you did not contribute input to?

12. Multiple drafts of the plan and schedule are discussed and recommended throughout this book. Have you ever written a school paper and submitted it without first printing a rough draft or, better yet, incorporating third-party review comments? What was the result?

13. Prepare a logic diagram utilizing the following list of schedule activities and their predecessors:

Activity	Predecessor
10	—
20	10
30	20
40	20
50	40
60	30, 40
70	50, 60
80	30
90	70, 80
100	90

Lean Construction Planning

5.1 INTRODUCTION

Preconstruction planning saves time, reduces costs, and improves quality and safety controls. Planning is defined as taking time before you act. Planning saves resources and is therefore also part of lean construction practices. Lean construction techniques, or in this case, lean planning, are a collection of processes intended to eliminate construction waste and still meet or exceed the project owner's expectations. The application of lean principles results in better utilization of resources, especially labor and materials. The strategy for lean supply is to provide materials when needed to reduce variation, eliminate waste, improve workflow, and increase coordination among construction trades. The project team's responsibility is to create a realistic construction flow that reflects the general contractor's (GC's) and the subcontractors' dependency on material suppliers. Material deliveries are then scheduled so that the materials arrive on site just as they are needed for installation on the project.

While lean planning does not replace a construction project's detailed contract schedule, it utilizes short interval planning and control that improves the timely completion of construction tasks. Another aspect of lean is the expanded use of off-site construction or prefabrication of building components. Reduced fabrication on the project site enables workers to concentrate on installation of material fabrications rather than creation of construction components, which ultimately saves time, reduces material waste, enhances on-site safety, and improves quality of installation. Acceptance and incorporation of lean for a company or project requires top-down support, from the ownership of the client, GC, and architect teams, all the way through design engineers and jobsite construction managers, to subcontractors and foremen.

Implementation of lean construction is the next logical step from activity-based costing (ABC) and other cost control methods, such as work packages and earned value, which are discussed in later chapters. Similar to ABC, there have been many academic articles and research publications on lean construction. In addition to ABC and lean planning techniques, this chapter introduces other advanced cost and time saving methods. This includes target value design (TVD), just-in-time planning (JIT), last planner, pull planning, value engineering (VE), subcontractor and supplier impacts, supply chain material management, and jobsite laydown and material handling.

5.2   ACTIVITY-BASED COSTING

The topic of activity-based costing (ABC) has been a focus of many cost engineers, researchers, and academics since the early 1990s. Activity-based costing is not necessarily a process to reduce costs, but more one of identifying indirect costs so that they can be applied to the work, allowing contractors to understand the true cost of direct construction activities. A study and application of ABC processes focuses on home office indirect costs, jobsite indirect costs, and fabrication facility costs. This will allow contractors to focus on improving costs and schedule where needed most and to specialize on those types of construction projects and construction activities which may be most profitable. Applying indirect costs to construction activities also allows a contractor to improve its estimating capabilities. It is generally understood that all contractors want to reduce costs and save time, and chapters on jobsite controls and earned value will later demonstrate some of the techniques that have been historically adopted. Controlling overhead costs has always been of interest to contractors, but the first step in ABC is to understand these costs and track them.

The purpose of activity-based costing is to track and assign as many of the home office and jobsite indirect costs as is possible to the work activities that they directly support. The ABC process begins with assigning costs to divisions within the company, construction projects, and ultimately, if possible, direct construction activities. There are many opportunities and benefits of ABC including identifying inefficiencies and unnecessary costs, specialization of types of projects, and strategic decisions on subcontractor or direct work packages. Activity-based costing identifies the more-profitable scopes and allows contractors to assign resources to those scopes and therefore improve profits. There are many potential ABC impacts to enhance accuracy of future estimates including an improved database incorporating as-built estimates, lowering prices of overpriced items (those which do not require much indirect support), and raising prices of underpriced items (those which require a disproportional amount of indirect support); this should improve pricing competitiveness in the marketplace.

This book introduces the new concept of *activity-based resourcing* (ABR). Where ABC assigns costs to the items of work such that the true cost of that item can be known, the concept of ABR is to assign resources to those same work activities. Resources are discussed throughout this book and include major cost and schedule categories, such as direct labor, material, construction equipment, and supervision and management personnel. Examples of these are the carpenter's time, supply of redi-mix concrete, and welding machine rental. Cash flow is also a necessary resource for contractors. The schedule is either prepared based on available resources or resources are applied to the schedule as needed and potentially balanced, as will be discussed later in this book. Every construction activity will require a mix of labor, material, equipment, subcontractors and other resources. Nonconstruction activities often are beyond the contractor's control, but still must be incorporated into the schedule. But activities that result in work on the jobsite will require resources, some of which may be assignable to activities, such as the carpenter's time for formwork and concrete supply for foundations. Other resources, such as a forklift or tower crane time, will be distributed over all work on the site and cannot be applied to any specific activity and therefore are not applicable to ABR.

5.3 LEAN CONSTRUCTION

There is not one exact definition of lean construction, but rather a body of research leading to a philosophy of planning and implementation of advanced cost and schedule control methods. Many lean planning topics are intermixed. Contractors do not need to adopt them all in order to accept a lean philosophy; in fact, some argue that a few of these concepts don't apply strictly to lean. But if "eliminate waste" is the ultimate goal, and loosely defined, then these all fit under the lean umbrella plus additional advanced jobsite control concepts; essentially there is no limit. The Lean Construction Institute's website defines lean as "enhances value on projects and uncovers wasted resources, such as time, movement, and human potential" (https://www.leanconstruction.org).

Like activity-based costing and other control topics, lean has been adapted to construction from production industries, such as the automobile industry and specifically from Toyota. Minimizing waste is not just for construction materials but eliminating any inefficiency (such as a surplus of inventory), improving labor productivity, and creating a satisfied client are all premises behind lean. Anything that does not add value must be eliminated; lean creates value. Lean is influenced by total quality management where the lack of quality in any facet of design or construction results in rework, which disrupts the flow of work. Increasing speed of design or construction in an effort to save money can actually cause waste and increase defects and therefore costs more money. The adage that "time is money" does not always hold true. Since construction starts with design, the root of many lean savings ideas stems from improving design operations. Although the term *lean construction* is widely understood, it could be expanded to include lean built environment, lean design, and lean planning by the GC's jobsite management team, which is discussed here.

Lean topics, such as pull planning or JIT, are not tools but processes. Lean maximizes value by minimizing waste. Lean is not happenchance; rather, it needs a formal process and adoption by all the team members for it to be a success. Acceptance of lean processes also requires a top-down commitment from design and construction company owners as well as project owners or clients. Repeat design- and construction- and project owner-teams adapt well to the lean process. If they have worked together, they know what to expect from one another. Lean is also best suited for complicated projects and projects that may use an integrated project delivery (IPD) approach, where all three parties share in the same risks and sign the same contract as shown in Figure 5.1. Integrated project delivery improves lean success and lean

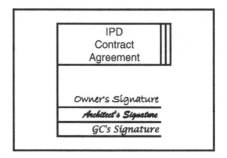

Figure 5.1 Integrated project delivery

improves IPD success. If IPD is not the project owner's chosen delivery method, then design-build or construction manager at-risk delivery methods would best implement the cost savings topics discussed in this chapter. Lean is also better suited for negotiated projects than bid projects.

The construction industry tends to resist new concepts, such as activity-based costing and lean, labeling them "production oriented." Construction contractors, with their proven means and methods, are also slow to change. It is common to hear out on the jobsite that "if it isn't broke, don't fix it." The construction industry is different from other industries, because no two projects are alike. Some identify that the problem with contractors is their detailed focus on independent activities, estimates, and schedules. Lean focuses on bigger picture goals and interaction of activities. Subcontractors also cause fragmentation in cost analysis and therefore make it more difficult for contractors to adopt lean or ABC. Many construction leaders, however, do have an open mind to improving their processes and have adopted many of the techniques discussed here.

The following subsections introduce target value design, just-in-time deliveries, last planners, pull planning, and supply chain material management. These concepts and processes are all subsets of lean construction. For this discussion, the term *lean planning* is synonymous with lean construction and is the preproject planning process to synthesize the construction team's means and methods, which will result in lean construction improvements and therefore time and cost enhancements.

Target Value Design

Target value design (TVD) is a subset of lean construction. TVD has also been referred to as target value costing and originated in manufacturing industries; again, the automobile industry in particular. Similar to lean, TVD works well with negotiated and design-build projects. The project owner sets the budget, and the design team is tasked to design to that budget. The budget is divided up among the different design disciplines like pieces of a pie. Each design package must financially fit within their piece of the pie, so the entire project meets the owner's budget; they design to

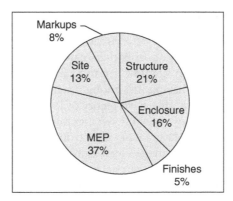

Figure 5.2 Target value design pie chart

the target cost. This process either assumes the designers have estimating capabilities in-house, an outside estimating consultant is engaged, or a construction manager is contracted early. Some critical subcontractors may need to be brought on to the team early during design development, such as mechanical, electrical, plumbing, precast concrete, curtain wall, landscape, elevator, and others. The estimate for the Fire Station 83 case study has been arranged in this fashion in Figure 5.2.

The owner's satisfaction is the number one priority for the design and construction team. The design team must anticipate cost and schedule when making design decisions rather than wait until bids are received. If the budget is exceeded the project must be redesigned, often through value engineering. The design must be targeted to "final" cost as a design parameter, which is often different from bid or contracted cost. Might this same concept be applied to scheduling? The TVD process includes active management of owner contingency logs and value-added logs, which are similar to a subcontract buyout log or value engineering log.

A target value design project requires collaboration between all of the parties, not only the project owner, architect, and general contractor, but also the design engineers and major subcontractors. The estimate is established from the top down, which in construction is the opposite from the way most estimators work, generally beginning with a more detailed focus, such as quantity take-offs and pricing recaps, and summarizing as they work toward the final price. In this regard, TVD is similar to pull planning, which is discussed further on, where the end date is determined and the plan is developed in reverse. If one element of the project is over budget, such as structural steel, money must be moved from another system to compensate, such as electrical. For a scheduling adaptation, could the end date be established, and the balance of the activities scheduled to meet that date? Like many lean concepts, TVD connects well with building information modeling. Fourth- and fifth-dimension modeling can help with early contractor-generated estimates and schedules.

Just-In-Time Deliveries

Just-in-time deliveries of construction materials has been adapted from just-in-time manufacturing. The goal is that materials which will be needed within 36 hours of installation are on site, but not too early or too late. The contractors perform a balancing act with having just enough material but not too much. Projects like the fire station case study that have plenty of laydown area do not have as great of need for JIT as would a downtown skyscraper but may still benefit from its principles. The City of Pasco is three hours east of metropolitan Seattle. Construction sites that are too large may not be managed lean as materials may be spread too far from their place of installation. Too much material on site for too long also exposes it to weather damage and potential theft. Doubling handling material is one of the most significant causes of labor inefficiency, as will be discussed further on in this chapter. But a problem with holding suppliers off from delivering too soon assumes that the fabricator has room in their storage yard (therefore not lean) and material can be stored out of the weather and secured. Other problems can occur when the fabricator sells this project's materials to another GC who needs it quicker or is willing to pay a premium and the supplier grabs whichever material is easiest to move or sell.

Last Planners

Planning is hard work; it is look-ahead scheduling. The individual or group of individuals responsible for accomplishing the detailed work is labeled the "last planners." This includes carpenter and electrician foremen on the construction side, and the structural and electrical engineers on the design side. The principal architect is not the last planner. Most architectural firms employ project managers just as do GCs, project owners, and subcontractors. The last planner concept should extend down through the organization chart to the project engineers, cost engineer, and scheduler if possible.

Many general contractors develop top-down schedules and dictate to subcontractors when their work is to be performed. A collaborative schedule created with subcontractor input, not by a marketing executive, but foremen and superintendents, is a much more feasible schedule. The same benefit can be had when a GC's proposed project superintendent assists with developing the estimate and the contract schedule.

Lean construction enthusiasts feel that contractors should not overcommit or overload their schedule. A little underloading is acceptable as that allows for adjustments during the process if necessary. Conversely many contractors would argue that crews need to be given just slightly more than what they can accomplish in a week, not less. Some contractors feel that employing a superintendent or foreman who is slightly underqualified to perform his or her duties but is ambitious and wants to work hard and learn, is a better fit than an overqualified individual. This philosophy would support a slight overload of work versus an underload. But supervisors should be cautious about too much overload or their foremen may become disenchanted as the next example reflects.

Example 5.1

Lean construction advocates indicate each foreman should have a little bit of leeway in their workload to accomplish new or changed conditions. This electrical foreman had historically utilized to-do lists for his short-term schedule, mostly prepared the afternoon before he went home from work. But this construction project had so many "fire drills" (i.e. new important tasks that had to be done each day) that he found he was struggling with his previous list of to-dos. He eventually lost interest in the to-do list and would leave it blank, knowing that the next day would fill up with new important tasks imposed on him by others. Unfortunately, priority items and longer-term planning were jeopardized, as was his enthusiasm to start each new work day.

Pull Planning

Pull planning or pull techniques are also adapted for lean construction from manufacturing industries. The people who are the pullers are the ones who use the output and would be considered the "last planners," as discussed earlier. The process may also begin with a lean planning workshop and facilitator. Each stakeholder should be represented, including subcontractors. Just as target value design starts with the total budget and works backwards, pull planning starts with work which needs to be in place in three to four weeks and plans backwards to accomplish that goal. Each contractor and designer is given a set of colored sticky notes and posts their delivery dates and commitments and what needs to happen to accomplish those dates. This would be comparable to each subcontractor developing a three-week look-ahead schedule, which is often done for the weekly foremen's meeting, and then those schedules would be shared with the other companies. But in this case, all of the companies collaboratively develop one schedule together. A photograph of a commercial project pull planning schedule is included as Figure 5.3.

The goal of pull planning is not only to achieve collaborative buy-in from all parties, but to work together to remove all constraints from accomplishing the work, so that the only thing left for the last planners is to get the job done. Some of the categories of potential restraints that should be considered include:

- Project owner has funded the project.
- Design is done and has been quality checked and change orders will be limited.
- Permits are in hand and inspectors are scheduled.
- Requests for information and submittals have been processed and answered.

Figure 5.3 Pull planning schedule

- Material is onsite or will be onsite.
- Qualified and sufficient labor is on hand.
- Equipment and tools are in working order.
- Predecessor activities have all been identified and will be accomplished.
- The jobsite is clean and safe.
- Sufficient time and cost estimates have been allowed for.
- No rework and no waste are the ultimate goals.

Many contractors were interviewed, along with scheduling consultants, when researching for this book. They provided input to draft tables of contents, and chapter drafts, and provided schedule examples for figures. When asked how they used pull planning and what technology tools were now employed the answers varied considerably and are reflected in the next example.

> **Example 5.2**
>
> 1. We still use sticky notes on a whiteboard.
> 2. We move all the tables and chairs out of the room and get subcontractors up to the board to place their preplanning notes.
> 3. We will be trying a new software on our next project (see Chapter 12).
> 4. We do not use pull planning. We have our subcontractors collaboratively input to the original detailed schedule and utilize three-week schedules in the field.
> 5. We use sticky notes on a large roll of butcher paper that is rolled up and pulled back out when needed.
> 6. We use "make-ready" planning, which includes sticky notes on a large calendar.
> 7. It depends on the sophistication of our superintendent and our trade partners.
> 8. Sometimes if we dictate to the subcontractors, they use us as their excuse but if we allow them too much input, we lose control.
> 9. Pull planning, or what we call "sticky-note scheduling," is required on every project. Even when the GC doesn't require it, we (subcontractors) nag for it until they finally relent.

5.4 VALUE ENGINEERING

Value engineering to some members of the built environment, especially some designers, implies cheapening of the project's design, but that is not true value engineering. Value engineering includes analyzing selected building components to seek creative ways of performing the same function as the original components at a lower life-cycle cost and/or faster without sacrificing reliability, performance, or maintainability. Value engineering studies may be performed by consultants during design development, as a contractor-performed preconstruction service, or by the GC during construction. The most effective time to conduct such studies is during design development. Target value design as just discussed needs to happen before design to be efficient; VE is the opposite and often happens either during the deign process or after design. The traditional lump sum delivery method of design – estimate – value engineering – redesign – building results in wasted time and additional design fees and therefore VE is not considered a lean construction topic by some.

Value engineering studies are typically performed during the preconstruction planning phase. The intent of the process is to select the highest value design components or systems. The essential functions of each component or system are studied to estimate the potential for value improvement. The VE study team needs to understand the rationale used by the designer in developing the design and the assumptions made in establishing design criteria and selecting materials and equipment. The intent of VE is to develop a list of alternative materials or components that might be used. Preliminary cost data is generated, and functional comparisons are made between the alternatives and the design components being studied. The intent is to determine which alternatives will meet the owner's requirements and provide additional value to the completed project in time and/or costs saved. Often life-cycle cost data is analyzed for each VE alternative before presentation to the project owner.

Preparation of value engineering proposals includes analysis of advantages and disadvantages of each alternative and the ones representing the best value are then selected for presentation to the designer and the owner. The VE proposal looks very similar to a post-contract change order proposal, including all detailed costs and markups and substantiation. All cost data prepared by the VE team should be included with the proposal, similar to a change order proposal. Each VE proposal is tracked in a log, similar to other schedule control or equipment tracking logs managed by the construction team. The proposals are submitted to the designer and the owner for approval. If approved, VE proposals should be incorporated into the contract documents. Value engineering proposals approved after the construction contract is awarded must be incorporated into the contract by change order.

5.5 SUBCONTRACTORS AND SUPPLIERS

Subcontracting is not necessarily a lean cost control topic, but it does have an impact on the general contractor's ability to implement activity-based costing and lean processes. The use of subcontractors by the GC is often a risk management choice. Subcontractors provide the GC with access to specialized skilled craftsmen and equipment that they may not have in-house. Craftspeople experienced in the many specialized trades required for major construction projects are expensive to hire and generally used on a project site only for limited periods of time. It would be cost-prohibitive for a GC to employ all types of skilled trades as a part of their own full-time workforce.

The general contractor has theoretically passed time and cost risk to its subcontractors, especially if they are contracted lump sum. In one sense, using the marketplace to buy out subcontractors and suppliers at the lowest competitive price is a cost reduction tool and is therefore lean but only if additional best-value attributes are considered beyond cost, such as quality, schedule, and safety performance. Ignoring the role of subcontractors in any discussion of lean planning is omitting 80–90% of the scope of the construction project. Subcontract management is the focus of Chapter 17 toward the end of this book. This brief introduction of subcontractors and suppliers naturally leads into material management covered in the next two sections.

5.6 SUPPLY CHAIN MATERIAL MANAGEMENT

Many lean construction research case studies assume the general contractor is also the material fabricator, which is not customary. General contractors rely heavily on support from subcontractors and suppliers to successfully build a project as discussed earlier. All construction materials are provided through suppliers. Even if the GC employs a subcontractor, the subcontractor will purchase materials from third-party suppliers. Cost and schedule control responsibility within the fabrication facility belongs to the supplier. Good supply chain material management by the GC also relies on shifting much of the on-site material fabrication and assembly to suppliers, especially local suppliers. Supply chain material management shifts the cost and schedule control responsibility as close to the beginning of a construction project as possible, to design and material fabrication, and not solely on the traditional focus of a contractor's direct jobsite field labor.

Off-Site Prefabrication

One process for improving construction productivity is the use of off-site construction and prefabrication of building components and modules. Off-site construction reduces material waste, improves worker safety, provides better quality control, reduces trade interference, improves worker productivity, and reduces on-site construction time. Expanded use of building information modeling has facilitated the increased use of off-site construction. Many electrical and mechanical systems are now prefabricated in specialized shops and delivered to project sites as assemblies instead of being constructed piece-by-piece on site. This eliminates requirements for craft labor to use ladders to perform work on the project site, improving productivity and safety. Use of off-site construction may affect the project owner's cash flow for the project because components are fabricated well in advance of their installation on site, so contractors need to ensure that owners are aware of the proposed payment schedule.

Removing or minimizing construction waste is a fundamental plank of lean construction planning and is accomplished through prefabrication of construction materials. Off-site prefabrication of materials improves *quality control*, as the materials are built and stored out of the weather, and utilizes in-process inspections; improves *safety* in a controlled environment; reduces *cost* as it changes fabrication of many construction materials from project-based to product-based; and improves *schedule* adherence as many activities can be performed in parallel rather than in series. Materials may be specified by the design team as purchased products, rather than field-assembled, but oftentimes a GC can weigh the advantages of shop fabricated versus field assembly as part of their preconstruction planning means and methods. There are many prefabrication shops and assembly yards that specialize in assembly and sale of construction materials. In addition to construction of single–family prefabricated homes, commercial construction examples of off-site material prefabrication include:

- Standard examples: Concrete rebar, precast concrete beams and planks, structural steel including joist and trusses, wood joist and trusses, cabinets, door assemblies, and mechanical ductwork.
- Occasional examples: Brick wall panels, apartment wall and floor wood-framed panels, structural insulated panels, bathroom plumbing carriages, and electrical cable tray.
- Potential examples: Concrete columns, concrete walls, elevated concrete floor slabs, and whole apartments or hotel units.
- The steel bow trusses for the fire station case study were fabricated off-site and efficiently erected in 75-foot-long assemblies.

Local Material Purchases

It is not only lean to purchase construction materials locally, but also sustainable and a good means of quality and schedule control. A GC that chooses suppliers that fabricate materials locally can easily personally visit the warehouse or yard and witness progress. Third-party inspectors may also visit the shop and inspect and test, as is the case with structural steel fabrication. Materials that are purchased and/or fabricated locally, say within 100 miles of the project site, will utilize less fuel in transportation. Purchasing locally results in additional Leadership in Energy and Environmental Design points or credits being awarded, if the project owner is pursuing a sustainable certificate or plaque. There is also a nonmeasurable, and maybe moral reason a contractor should do business with as many of its local neighbors as is feasible.

5.7 JOBSITE LAYDOWN AND MATERIAL HANDLING

It is imperative that the jobsite be organized efficiently in order for the construction project to be successful. A jobsite laydown plan is a useful tool to manage material receipt, handling, and storage. The site layout plan should be developed by the general contractor's superintendent and it should consider not only the needs of the GC's direct work, but also the requirements of all of the subcontractors working on site. The site plan should indicate:

- Existing site conditions, such as streets, adjacent buildings, and overhead and underground utilities.
- Planned locations for all temporary facilities, such as fences and gates, trailers, temporary utilities, sanitary facilities, erosion control, and drainage.
- Material handling and storage areas, equipment storage and access, craftsmen and visitor parking, and hoisting methods, such as a tower crane.

The superintendent should consider all site constraints, equipment constraints, jobsite productivity, material handling, and safety issues when preparing the jobsite layout plan. Some additional factors that should be considered in developing the jobsite plan include traffic flow on and off the site and location of site offices convenient for visitor control. Some of the goals of the site logistics plan are to: Eliminate double handling of material which is a lean productivity factor, protect materials from weather and theft, and keep materials close to the installation place but not so close that the craftsmen are stumbling over them. These are all JIT lean considerations as was discussed previously.

Equipment needs for material handling and hoisting should be reflected in the jobsite plan. The proper choice of tower crane types and locations is essential for a productive plan, especially on downtown hi-rise projects. The fire station case study project fortunately had 4.7 acres of site with plenty of laydown area and access for mobile cranes, although the contractor utilized only a little over one acre of the available space.

In addition to a productivity tool, the jobsite layout plan is a great proposal, interview, and marketing tool. It shows the project owner that the contractor has thought through the project – a personal touch. It may make a difference on a close award decision with a negotiated project. The jobsite layout plan for the case study is included on the book's website.

5.8 SCHEDULING LEAN

Lean construction processes or techniques are typically focused on cost savings and reduction in waste as reflected in this chapter. But all of these lean processes can also be thought of as time savings – it is a double win as reflected below.

- Activity-based costing or scheduling: By assigning the actual time it takes to install an item of work, beginning with unloading it off the truck, storing and handling, installing, testing, inspecting, and potentially punch list, the scheduler will know the true duration of an activity.

- Activity-based resourcing: If resources are applied to schedule activities, rather than uniformly distributing them across all work or the entire duration, each activity's use of resources can be found out and utilized for future planning and resource balancing efforts.

- Target value design: Instead of dividing the TVD pie up according to budget, it could also be divided up with respect to time. Each system is allowed so much time, or portion of the project duration, and if any system like glazing needs more time, then another system, such as roofing, needs to be expedited so that the entire project duration can be accomplished.

- Just-in-time deliveries: Material deliveries received too early must be moved around the site, which takes resources (reference ABR), and materials that are received too late could potentially impact the project schedule. Administrative and nonconstruction activities, such as submittals, fabrication, and deliveries, must be shown on the schedule, otherwise there may be errors in the planning logic.

- Last planner: The people doing the work, such as the foremen and subcontractors, should be developing short-interval schedules. The project superintendent should have been involved in the initial project schedule development and he or she reports on schedule status at the weekly owner-architect-contractor meeting.

- Pull planning: One method to prepare short-interval schedules is to have the last planners schedule backwards, from known milestones or constraints, and plan the work for the upcoming three weeks prior so that the milestone can be achieved.

- Value engineering: It is customary to think of costs saved by the introduction of VE alternatives, but the cost trade-off might be neutral or even an increase in cost, if the time savings associated with a VE proposal is beneficial to the project. A VE proposal may also be beneficial to a project if it improves schedule, even if there might be a slight increase in cost.

- Subcontractors and suppliers: Without input from 80–90% of the workforce in development and control of the project schedule, the GC cannot be successful. Collaborative involvement from these last-planners and team members is crucial to schedule success.

- Off-site prefabrication of materials: Many more activities can occur in parallel rather than series due to off-site prefabrication. Just-in-time deliveries also enhance schedule adherence and reduce chances for on-site material theft and damage.

- Local material purchases: This is good for the environment due to less fuel needed for material deliveries, but it is also good for local businesses when contractors "buy local." And if materials are fabricated near the site, then tighter coordination for delivery times, or JIT deliveries, will also benefit the project schedule.

- Jobsite laydown: The preparation of a jobsite laydown plan by the project superintendent will result in less double-handling and material damage and enhances ABR as mentioned above.

5.9 SUMMARY

Lean construction planning, loosely defined, includes designing, fabricating, and building construction projects as cost effectively as possible, eliminating waste, all the while maintaining the same if not improved levels of quality, schedule, and safety control. Lean construction has several subcategories including target value design, just-in-time material deliveries, last planner participation, and pull planning scheduling processes.

Activity-based costing is a pre-cursor to lean construction. Before costs can be managed or reduced, it is important to know which activities are driving the costs and the total costs associated with each activity, including jobsite indirect costs and potentially home office overhead.

Target value design starts with the project owner setting a strict budget that both the design and construction teams must adhere to. Construction projects, such as skyscrapers, have been practicing just-in-time material deliveries before JIT was included as a study of lean construction. Efficient JIT management requires delivery of materials when needed, not too early so that double handling is minimized and so late that it affects the installation schedule.

Ultimately the people involved in designing and installing materials, the last planners, should be the ones involved in estimating, scheduling, and cost control. Essentially this is the structural engineer who sizes the spot footing and the carpenter foreman responsible for excavation, forming, and placing the footing concrete. Pull planning is a collaborative scheduling approach that includes those last planners preparing short-interval schedules by scheduling backwards from given milestones, often with multicolored sticky notes.

Value engineering seeks the same goal as target value design but happens generally after some design has been completed. Contractors prepare cost saving or life-cycle saving proposals for project owners and designers to consider and accept, which then require redesign and incorporation into the GC's contract and various subcontract agreements. Subcontractors comprise 80–90% of the workforce on any typical construction project. True lean planning should focus on subcontractors and suppliers.

Supply chain material management refocuses cost and schedule control from onsite labor to the source of material design and fabrication. This is a more holistic approach to lean construction, starting at the beginning and chasing down material sources, including delivery times and cost of fabrication. Shifting site labor to off-site warehouses and supplier fabrication yards saves field labor and potentially improves quality, safety, and schedule control. Some material examples which benefit from off-site prefabrication include structural components, such as wood and steel trusses, concrete beams and columns, and finish materials, such as cabinetry and door assemblies. Off-site prefabrication also reinforces JIT planning. It is not only good for business and community economies to purchase materials from local material suppliers and employ local subcontractors, but it is good for the environment due to reduced fossil fuel use in material deliveries. Purchasing local also allows the GC and project owner and design team to visit fabrication shops and perform early quality control reviews. Sourcing local also has safety enhancement considerations and employs JIT material planning. Local material purchasing employs many lean and sustainability planning and construction techniques.

A preconstruction jobsite laydown plan should be developed by the project superintendent. This plan, which includes many of the lean construction planning processes discussed in this chapter, is a physical drawing of the project site. The superintendent will include many cost, schedule, quality, and safety enhancement considerations, such as material laydown, stormwater control, traffic routes, and others. All of these are advanced cost and schedule control methods and should be considered part of lean construction planning. There are many more as well, and others yet to be discovered and implemented by future construction managers.

5.10 REVIEW QUESTIONS

1. What is the difference between lean construction and activity-based costing?
2. What is the difference between target value design and value engineering?
3. What is the difference between a subcontractor and a supplier?
4. What is an advantage of having too much material on site versus not enough material?
5. If room is not available to store material on site, where can it be stored?
6. What happens if the doors, door frames, and door hardware package is bid higher than what was budgeted on a TVD project?
7. List three advantages of off-site prefabrication versus on site casting for a system, such as concrete beams.
8. What is the difference between a postcontract VE proposal and a change order proposal?
9. If a project has unlimited laydown area, why might it still be advantageous to employ JIT techniques?
10. Who should develop the jobsite laydown plan and why him or her?

5.11 EXERCISES

1. Other than the examples listed in this chapter, what types of materials or material systems might be prefabricated off-site?
2. Prepare an argument regarding why it would be better to (A) schedule your foreman for only 90% of his or her capacity, and/or conversely (B) schedule your foreman for 110% of his or her capacity.
3. Do you know of any other lean planning techniques?
4. Prepare three value engineering ideas from your case study and include them in a VE log.
5. Other than cost and schedule considerations, why might it be advantageous for a contractor to purchase materials fabricated locally and employ local craftsmen and subcontractors?
6. Other than a hospital, list three project types that may be suited for lean construction processes.
7. The fire station GC had 4.7 acres of site available but only utilized a little over one acre for laydown. Why was that?
8. List 10 items that should be incorporated into a jobsite laydown plan.
9. Develop a pull planning schedule utilizing sticky notes for the next three weeks. Use different colors for different classes, work, and social activities. Identify milestones or constraint dates and incorporate activities necessary to accomplish them.

Chapter **6**

Contract and Time Considerations

6.1 INTRODUCTION

There are many important construction documents but the prime contract agreement itself is *the most important* of all construction documents. Contracts are adequately covered in many construction management (CM) books (e.g. *Management of Construction Projects, a Constructor's Perspective*), so that information will not all be repeated here. This chapter includes a brief introduction of contract considerations, especially as they relate to time and scheduling. Before project owners offer a contract to a general construction (GC) firm, they must first decide their procurement and delivery methods, as introduced in Chapter 2.

This chapter begins with the documents that make up the contract agreement, which are also known as contract documents. The contractor's front office, including the scheduler, must make note of the specific contract language and whether the owner intends for the schedule itself to be added as a contract document or exhibit. In addition to the construction schedule, many contract clauses have an element of time, such as pay requests, termination clauses, notice clauses, claim processing, retention release, and others. Also of importance is the format an owner may dictate the schedule take as well as inclusion of hard constraint dates and/or durations in the contract agreement. Several contract documents, special conditions to the contract, and technical specifications were reviewed in preparation of this book, and this chapter includes several time-related contract references.

All construction contracts present risks to the general contractor, and the project manager (PM) and superintendent need to develop a risk management plan for each project. The allocation of risks between the project owner and the contractor is defined by the terms and conditions of the contract. Risk identification and management is a critical jobsite management function and is also discussed in this chapter.

6.2 CONTRACT DOCUMENTS

The construction contract is the most important construction document. It has significant impacts on how the schedule is prepared and maintained. The contract is a legal document that describes the rights and responsibilities of the parties, e.g.

the owner and the CM or GC. Five things must be aligned for a contract to exist. These include:

1. An offer to perform a service;
2. An acceptance of the offer;
3. Some conveyance or exchange, such as the transfer of a completed building for money;
4. The agreement has to be legal; and
5. Only authorized parties can sign the agreement, for example, the officer-in-charge or chief executive officer.

The intent of the contractual agreement and the contract documents describe the completed project and the terms and conditions the parties (usually the project owner and the GC) must adhere to in order to accomplish the work. There typically is not a detailed description of the sequence of work or the means and methods to be used by the contractor in completing the project. The contractor is expected to have the professional expertise required to understand the contract documents and select appropriate subcontractors and qualified craftsmen, materials, and equipment to complete the project safely and achieve the specified quality requirements. For example, the contract documents will specify the dimension and sizes of structural steel columns and beams, but will not provide the design for erection aids, such as erection bolts or erection seats or hoisting and safety considerations. The contract may stipulate a start and completion date, but not how the contractor achieves those dates. The contract documents usually include at least the following five essential elements:

1. The contract itself or prime agreement,
2. Special or supplemental conditions,
3. General conditions,
4. Technical specifications, and
5. Drawings.

The *prime agreement* describes the project to be constructed; the pricing method to be used and cost; the time allowed for construction; and the names and points of contact of the project owner and the contractor. Contractors and project owners use a variety of risk management strategies including insurance and bonds. The requirements for these will also be included in the prime contract agreement. Supporting documents may be incorporated into the agreement by reference or as exhibits; these documents then also become part of the contract documents. Many additional potential contract documents are listed below.

Some owners use a *project manual* that contains the special conditions, general conditions, soils report, erosion control requirements, prevailing wage rates, technical specifications, and other project-related documents. The invitation to bid and request for proposal are generally not considered contract documents unless referred to in the prime agreement or included in the project manual.

Special or *supplemental conditions* are customized for each specific project. They include issues such as work times, liquidated damages (LDs), site and parking conditions and other unique requirements. Special conditions may be included in the project manual or as part of Construction Specification Institute (CSI) divisions 00 or 01 within the technical specification book. The *general conditions* provide a set of operating procedures that the owner typically uses on all projects. They describe the relationship between the owner and the contractor, the authority of the owner's representative or agent, and the terms of the contract. Some owners use standard general conditions published by professional organizations, such as the American Institute of Architects (AIA). A popular set of general conditions that accompanies many prime contracts is the AIA A201. Many requirements are contained in the general conditions but project-specific requirements are defined in the special conditions of the contract.

The *technical specifications* provide the qualitative requirements for construction materials, equipment to be installed, and workmanship. The old 16 CSI divisions have been replaced with a new listing of 49 divisions, but construction managers will find examples of each being utilized today. The *contract drawings* show the quantitative requirements for the project and how the various components go together to form the completed project. The drawings represent final placement and configuration of construction materials and systems, but not how the work is to be accomplished.

The American Institute of Architects previously established a system of precedents within these five document types in that if there was a conflict, typically the most detailed or specific document would govern. The drawings and general conditions were the least detailed and the contract agreement itself was the ultimate authority. Today the AIA advises that all of these documents work together and be complementary. In this way, a contractor cannot seek out discrepancies and use pre-established precedents to generate change orders. If any item of work is shown in any of the documents, it is included in the scope of work.

Contracts are either standard or specially prepared agreements. Larger government agencies have developed their own construction contract documents. Federal and state agencies typically have standardized general conditions and agreements. Many local government agencies and private owners use contract formats developed by the AIA or the new family of construction contracts from ConsensusDocs headed up by the Associated General Contractors of America (AGC). Contracts should not be signed until they have been subjected to a thorough legal review. This is to ensure that the documents are legally enforceable in the event of a disagreement and that there is a clear, legal description of not only the work to be performed but each of the contracting party's roles and responsibilities. The advantage of using standard copyrighted contract forms is that they have been developed by those skilled at contract law and have been tested in and out of the legal system.

Potential Contract Documents

Virtually any document can become a contract document, including notes on a cocktail napkin – although it is not recommended any of the parties use cocktail napkins for construction contracts. Some of these potential documents are more likely than others to make it into a construction contract. Documents that are included or referenced within the bid documents, such as within a project manual, and the bid documents themselves are typically included in the contract by reference. Other documents are added to the contract during the bid cycle or before the contract is executed if they are attached to addenda. Documents can be added to the contract after its execution and throughout construction if they are attached and/or referenced within agreed change orders. Addenda and change orders often have schedule clauses. Some of the additional contract documents that may have schedule considerations include:

- Request for proposal,
- Instructions for bidders or request for quotation,
- Letter of intent and/or notice to proceed, and
- Preconstruction agreement.

It is imperative that the contractor review all of these carefully before submitting a bid or finalizing the contract schedule. There are an infinite number of documents that would not typically be found in the original contractual agreement but any may be added later by reference or attachment to a change order that is negotiated into the contract. Some of these may also affect the schedule. All of the documents that comprise the contract are shown in Figure 6.1.

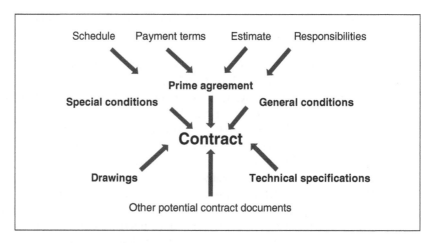

Figure 6.1   Contract document structure

6.3 CONTRACT LANGUAGE

The terms and conditions of the relationship between the primary parties are defined solely within the contract documents. These documents should be read and completely understood by the contractor before deciding to pursue a project. They also form the basis for creating a project estimate and schedule. To manage a project successfully, the GC's PM must understand the organization of the contract documents and the contractual requirements for his or her project. This knowledge is essential if the jobsite team has any expectation of satisfying the goals of their company executives as well as that of the project owner.

Most of the typical copyrighted contracts contain specific language regarding schedule. The contract must be completely understood before engaging in a bid or proposal process and certainly before executing the agreement. This includes not only the prime agreement but all of the referenced documents and exhibits. Scheduling receives a lot of attention in contracts. The following documents have been utilized for this research and examples of typical contract language that the scheduler must make note of have been paraphrased here. A couple of them have been introduced in previous chapters.

- American Institute of Architects A101, The Standard Form of Agreement between the Owner and Contractor, where the basis of payment is stipulated lump sum; authored by architects and generally favors the position of the design and owner teams; this and the A102 both reference the A201 General Conditions throughout, including a note on the first page that "A201 is adopted by reference"; A101 is only six pages in length.
- AIA A102, Standard Form of Agreement between the Owner and Contractor where the basis of payment is the cost of the work plus a fee with a Guaranteed Maximum Price (GMP); 15 pages in length.
- AIA A201, General Conditions of the Contract for Construction; generally, all AIA contracts reference this document; 50 pages in length.
- ConsensusDocs 500, Standard Agreement and General Conditions between Owner and Construction Manager, where the Construction Manager is at-risk; the general conditions and the prime agreement are merged into one document – this is essentially the same as adding the A102 to the A201. The ConsensusDocs family of contracts was developed by the AGC and a consortium of other contractor associations including specialty contractor organizations. The language therefore favors contractors over the designer and owner; in fact in many instances the designer is not included in specific contract clauses as noted elsewhere in this book; 43 pages in length.
- ConsensusDocs 751, Standard Short-Form Agreement between the Constructor and a Subcontractor; ten pages in length.
- Fire Station 83 Special Conditions to the Contract and Technical Specifications; primary case study for this book; the complete specification book is over 1,000 pages in length.

Contractual Terms

There are many terms, clauses, or phrases that the contractor's team, including the scheduler, must thoroughly research within the proposed contract documents that affect their approach to scheduling, and potentially fee and contingency decisions. Some of these terms are included here and others are included with Chapter 18, "Schedule Impacts."

- Liquidated damages,
- Calendar dates,
- Constraint dates, including start and finish dates and other milestones,
- Notice requirements,
- Force majeure,
- No claim for delay,
- Requirement for updates or revisions,
- Requirement for schedule format,
- Payment terms,
- Retention held and release of retention requirements, and
- Commencement or day one of the schedule.

Liquidated Damages

The schedule has important contractual implications and should be referenced and incorporated into the prime contract and each subcontract as a contract exhibit. This is not always allowed on public projects. Liquidated damages may be imposed if a contractor finishes the project late. If LDs are required, they must be clearly spelled out in all contractual agreements. In lieu of liquidated damages, some owners prefer actual damages. Conversely, some contractors encourage adding LDs into the contract. The following two examples reflect different approaches to LDs.

Example 6.1

Most contractors avoid a liquidated damages clause in their contract, but this particular GC insisted on inserting one, even on small negotiated projects – which was their specialty. Their theory was that if the contract had a LDs clause, even at a very small amount, such as $100/day, this was better than a 'real' or consequential damages clause that they would explicitly exclude.

Example 6.2

A hotel near the airport had to be complete and opened by a certain date to support a large convention, so the project owner included a $20,000 per day LDs clause into the contract. The GC asked for an equal amount of bonus if they finished early, but the hotel felt that was too steep and they settled at a bonus of 50% of the damages, or $10,000 per day. This 50% bonus to penalty ratio is common. The contract utilized a GMP pricing method. The GC developed savings during buyout and concrete construction and used these savings to pay the subcontractors to work selective overtime to get ahead of schedule, including working weekends. The GC also developed a unique tower crane (TC) approach. The project was a long, narrow, tall hotel and one TC could not reach from end to end. Competing GCs figured two expensive TCs in their jobsite general conditions estimates. But this GC put the TC on rails, similar to a railroad, and moved it during the night, from one end of the building to the other. The contractor finished the project one month ahead of schedule and asked for its 30-day or $300,000 bonus. The hotel responded that they had no intention of enforcing the LD clause, it was just to get the contractor's attention and therefore they had not intended to pay a bonus. The two parties settled on a smaller bonus.

The following contract clauses specifically address LDs:

- AIA A101, specific for the fire station, indicates in Article 4.5 that "Liquidated damages shall be in the amount of $1,070 per calendar day beyond substantial completion for the Fire Station 83 project."
- AIA A102 Article 5.1.6 refers to liquidated damages, "if any" . . . LDs are more likely associated with competitively lump sum bid than negotiated GMP projects.
- The A201 general conditions reinforces in Article 8.2.4 that "The owner will incur serious and substantial damages if completion does not occur within the contract time."
- ConsensusDocs 500 Article 6.6 discusses liquidated damages and allows the parties to choose between two boxes if the project "shall" or "shall not" impose liquidated damages, and if "shall" is checked, the article proceeds with "The construction manager agrees if the date of substantial completion (and/or another separate paragraph for final completion) is not attained, the construction manager shall pay the owner $500 per day as liquidated damages."
- ConsensusDocs 751 Subcontract Agreement, Article 7.3, references the prime agreement for terms if there are liquidated damages. It adds that both of these parties "waive all claims of consequential damages between each other." In this instance, whatever terms the owner ties the GC with, the GC passes along to its subcontractors, which is a standard practice.

Float

Some contracts indicate the owner and general contractor are to share the float equally, although this appears to leave the design team and the subcontractors out of the equation. The Army Corp of Engineers indicates that the float is not exclusive to one party or the other. If the contract is silent with respect to float ownership, both parties assume they own the float. None of the five contracts just discussed, or the fire station special conditions, addressed float. The concept of schedule float will be further discussed in Chapters 9 and 14.

6.4 SCHEDULE INCLUSION

The schedule should be referenced in the prime contract as an exhibit. It should also be referred to in each subcontract and purchase order (PO) by title and date. Unfortunately, the AIA A101, A102, and ConsensusDocs 500 do not include the schedule as a standard contract document, but each has an opening for "other documents" toward the end of the agreement, which a contractor may choose to utilize by adding the contract schedule. Although not specifically listed as a contract document, the contracts still require the GC to prepare and submit a schedule but with slightly different language as indicated in these articles:

- AIA A201 general conditions, Article 3.10, titled "Contractor's Construction Schedules" includes (parenthesis added for emphasis) "Submit for owner and architect's information (not approval) and revise as required . . . contractor shall prepare a submittal schedule . . . submit to architect for approval (not information), must be coordinated with the contractor's construction schedule . . ."
- ConsensusDocs 500, Article 6.2.1 indicates owner activities are to be included in the schedule . . . "The construction manager shall submit to the owner a schedule of the work including dates that information and approvals are required of the owner."

Subcontract Agreements

The contract schedule should be referenced in the prime contract agreement as an exhibit, whether it is developed as a top-down schedule or collaboratively prepared as discussed throughout this book. It should also be referred to in each subcontract and PO by title and date. The ConsensusDocs 751 was developed by GCs, CMs and subcontractor associations together and is a collaborative document in many regards as noted specifically for schedule in the articles listed here. Additional references to this contract will be made in other chapters.

- Article 3 allows for inclusion of the schedule as a contract document, specifically Exhibit C.
- Article 7.1 indicates "Time is of the essence for both parties" . . . not just one.

- Article 7.2 indicates the schedule development and maintenance is a collaborative effort between the constructor and the subcontractor as they both work together with progress and updates.

A general contractor will endeavor to pass schedule risk down to its subcontractors, including any exposure to liquidated damages as noted above. The GC passes similar risks and responsibilities outlined in its contract down to its subcontractors in regards to insurance and bond types and amounts, retention held, requirements for retention release, and pay request timing. This is typical regardless if contracts were copyrighted by the AIA, ConsensusDocs, or home-grown. The GC's PM may insert language into all subcontracts, placing the subcontractors on notice that they are required to achieve the project schedule and that if the GC determines that they are falling behind, the subcontractors will work overtime and/or increase crew sizes at their expense to catch up. An example of this is included in Chapter 17 that expounds on schedule requirements in subcontract agreements.

6.5 CONTRACTUAL SCHEDULE FORMAT

The contract may require the contractor to produce a schedule on a prescribed format or with specific software and may limit activity durations to no more than 15, 20, or 30 days or activities with a monetary value no greater than $50,000. The Fire Station 83 special conditions to the specifications included the following sections related to schedule format:

- Specification section 013200, Construction Progress Documentation, paragraph 1.3 included several schedule definitions including activity, float, and others.
- Paragraph 2.1 required the contractor to "submit a schedule of submittals in chronological order by the dates to be submitted," which is difficult for the contractor to do this early in the project.
- Paragraph 2.2 indicated the "contractor's construction schedule must comply with AGC's *Construction Planning and Scheduling* as relates to time-frame, activities (no longer than 14 days), including submittal processing, constraints, milestones, and others."
- Sub paragraph F requires the contractor to "Choose between two schedule software methods: MS Project and Primavera.

Some advocate the contractor prepare two schedules, one for presentation to the project owner and design team and updated to reflect work performed at the weekly owner-architect-contractor meeting, and another 'working' schedule that the superintendent builds to. There are complications and potential repercussions with this approach. The contract schedule is one of the first documents used to substantiate a claim for additional time and/or money and if it is discovered that two "sets of books," or in this case two schedules, are being used in parallel, they will both lose credibility.

The Fire Station 83 special conditions, section 013200, paragraph 2.3, titled "Critical Path Method" also requires the GC to "Submit a diagram within 14 days of issuance of a notice to proceed . . . including a cash requirement prediction based on indicated activities . . ." This is then essentially a cost-loaded schedule, which is discussed in detail in Chapter 11. This article includes many detailed actions for how the contractor is to prepare and manage the schedule, including float, schedule updates, early and late start and completion dates. Many of these directions are disconnected and it has likely been borrowed from another source or two. Paragraph 3.1 of the same section, titled "Execution of the schedule," further discusses schedule updates, baseline schedule, revisions to the schedule, reports, and schedule progress.

6.6 CONTRACTUAL TIMELINE

Commencement

Day one on the schedule may be as simple as the owner issuing a notice to proceed (NTP), which is not necessarily in the contractor's favor. Conversely, the project owner may require several documents before the contractor is allowed to proceed, for example performance and payment bonds, schedule of values for billing purposes, lists of subcontractors, quality and safety plans, certificate of insurance, and a detailed construction schedule. But receipt of only an NTP may not be in the contractor's favor, as the owner may not have yet obtained financing or building permits, or the parties may not have executed their contract. If the GC proceeds based only on an NTP, and they have a fixed duration or fixed end date with potential liquidated damages, they will not be in a strong negotiating position regarding schedule with the owner.

Signing a contract that only specifies a date, such as June 9, 2022, as the contractual start date, results in the contractor accepting the schedule risk. Therefore, many GC PMs will want to insert additional conditions into the contract, rather than just a date. These conditions would typically include receipt of construction financing verification, building permits, signed contract, and the NTP. Most project owners and designers would rather choose to insert a specific date in both the commencement and completion articles of the construction contract. Some contracts may specify start and/or end dates and ignore durations, and others just include durations.

The start date of different contracts often defaults to the notice to proceed or contract date, unless a specific date or other requirements are inserted. Some additional specific contract language relating to commencement includes the following:

- AIA A101 Article 3, titled "Date of commencement and substantial completion" . . .. Specific for the Fire Station 83 project indicates "Work shall commence as stated in the owner's notice to proceed . . ."
- Consensus 500 Article 6.1 states "The date of commencement is the agreement date unless otherwise set forth below" and allows for prerequisites as indicated earlier.

Duration

Contract language specific to durations of construction is similar with many of the contracts. ConsensusDocs 500 Article 2.3.5 defines contract time as the time between the date of commencement and final completion. Other contracts are similar except they may substitute "substantial" completion for "final" completion. ConsensusDocs Article 6.1.1 adds "The date of substantial completion shall be established in a separate Amendment 1 to this agreement." This amendment could be a schedule exhibit. Potential schedule delays and subsequent claims are discussed in Chapter 18.

Work Days

Another point of contention relates to the units used for *project duration*. Contractors usually prefer that completion is defined in a quantity of *work days* after the issuance of the NTP and the other aforementioned requirements. On the other hand, project owners prefer using *calendar days* jointly with specified start and/or completion dates, which can burden the GC if any of the requirements were not accomplished.

From a constructor's perspective, the schedule plot should use work days and not calendar days. The AIA A101 and A102 contracts, Articles 3.3.1 and 4.3, respectively, and other contracts will default to calendar days unless the contractor changes it when negotiating the terms. Calendar days include seven days a week and 365 days a year. Calendar days include Saturdays and Sundays, holidays like Christmas, and days that cannot be worked due to inclement weather conditions. Work days for commercial contractors are usually Monday through Friday, eight hours a day, a maximum of 40 hours a week, and not working holidays. Any time worked beyond that and the contractor has to pay its craftsmen overtime. Industrial projects and heavy civil projects with shutdowns will routinely require evening and off-shift work.

When is a day not a day? Not all contract language agrees, as indicated in the following:

- AIA contracts default to calendar days unless the agreement is modified.
- AIA A201 general conditions devotes nine paragraphs in Article 8 for "Time," including paragraph 8.1.4 which defines "day" as a "calendar day."
- ConsensusDocs 500 Article 2.3.2 regarding definitions indicates that business days excludes weekends and holidays, but 2.3.9 slightly conflicts by defining "day" as a calendar day.

Completion

Many construction projects have to achieve a hard completion date, such as schools opening in September, highway projects before 5:00 a.m. on Monday, the opening of large retail box stores before holiday shopping, a professional athletic stadium before the first game of the season, and others. Just as commencement, durations, and work days differ with various contract formats, so does completion as reflected here.

- AIA A101, Article 3, and A102 Article 4.3 share similar language (parenthesis added for emphasis): "The contractor shall achieve substantial completion of the entire work (and one of two boxes is to be chosen) . . . not later than *365* (for the Fire Station 83 project) calendar days after commencement or . . . by the following date –__." A102 includes similar language in Article 4.3.1.
- ConsensusDocs 500 Article 6.1.1 is very loose with the following language: "Unless the parties agree otherwise, the date of substantial completion or the date of final completion shall be established in Amendment 1 to this agreement . . ."

Occupancy Considerations

There are also many potential occupancy considerations that the contractor's scheduler should make note of when reviewing the contract. These can affect not only the schedule, but also cost, especially general conditions cost and time to accomplish contract close-out, which is necessary to trigger release of retention. There are two primary *certificates of completion* that the contractor has their eye on, including a *certificate of occupancy* (C of O) issued by the city that allows the owner use of the building, and a certificate of substantial completion which is issued by the architect which indicates the building is complete to the degree (not necessarily 100%) that the owner can use it. The C of O indicates the building is safe from a fire and life safety perspective and has been built according to code and the terms of the permit. The local fire department conducts one of the final inspections and their buy-off is critical to receipt of a C of O. A temporary certificate of occupancy (TCO) may be issued by the city to allow partial use of the project while the balance (such as landscaping or tenant improvements) is remaining. The TCO typically has an expiration, such as six months, before the final C of O needs to be achieved.

The *certificate of substantial completion* is often issued with an attached punch list. This certificate does not address code or egress issues as much as the C of O does, but rather whether the building is complete enough for the owner's use. Receipt of the certificate of substantial completion from the architect typically signifies the end of a contractor's exposure to LDs and the commencement of the warranty period.

The contractor needs to receive both certificates before they are complete and the owner needs to receive the C of O before they can use the project. Receipt of all of these completion certificates is one of the most important scheduling goals of any GC project superintendent. The contractor expects to receive its final release of retention, which approximately equals its fee, within 30 days of completion and that release is typically conditioned upon issuance of these certificates.

The ConsensusDocs 500 contract refers to final completion and portions of Article 2.3 state in part (parenthesis added for emphasis) . . . "Final completion is established by issuance of the certificate of final completion signed by the owner and construction manager . . . (note: not the architect) . . . and the date of this certificate . . . is the date when the owner may use the project," and Article 10.6 adds "The construction manager will notify the owner when substantial completion is met and

issues a certificate of substantial completion at which time warranties will commence
. . ." (note again: not issued by the architect). Article 10.8 further discusses "Final
completion and final payment which occurs 20 days after the construction manager
submits its application for final payment."

The ConsensusDocs contracts differ from several articles and paragraphs within
the AIA contracts where the architect establishes the certificate of substantial completion date and authorizes final payment to be released. Specifically, A201 includes several paragraphs in Article 9 which read in part: ". . . so the owner can occupy . . . the
architect will prepare a certificate of substantial completion at which time warranties
will commence.. . ." Issuance of final completion and final payment also has many
prerequisites.

It is understandable that the project owner is anxious to get into their new building and begin utilizing it as soon as feasible. They have spent years planning the
project and have a substantial amount of money invested in it. Sometimes the owner
cannot understand why they cannot use the building earlier, since "they paid for it."
Some related contract clauses include: Beneficial occupancy, dual occupancy, early
occupancy, partial occupancy, joint occupancy, and others. If the owner takes over
some use of the building, while the contractor is still working or on-site, then questions come up over cost and responsibility for a variety of areas. The best way to
resolve these is to discuss them long before the joint occupancy occurs. Some of the
potential problems include:

- Temporary utilities, such as water and power – who pays?
- Dumpster service – who pays?
- Warranty – starts when?
- Unions – if the GC is union, but the owner is not and the owner's vendors are
  working in the building, how is this resolved?
- Ongoing cleanup – whose garbage is it?
- Safety for contractor and owner employees and guests – if someone gets hurt,
  whose fault is it?
- Security for the building and the construction tools and materials – is this
  required and if so, who pays?
- Insurance – who pays?
- Punch list versus maintenance versus warranty repairs – from which bucket are
  these paid?

Many contracts identify partial or dual occupancy dates, such as ConsensusDocs
500 Article 10.7, AIA A201 Article 9.9, and others. Resolution of the above questions, however, is rarely addressed in any formal agreement but should be communicated between the parties if early or joint occupancy is a goal. Close-out is not
specifically addressed in this scheduling book but final close-out is a critical milestone
that receipt of a contractor's retention check is often linked to. Achievement of final

close-out requires resolution of all of the issues discussed here, along with submission of many close-out documents, all of which should also be specifically addressed in the contract agreement.

6.7 RISK ANALYSIS

As discussed throughout this book, construction is a risky business, and a construction contract is a mechanism for transferring some project risks from the owner to the contractor in return for some type of payment. The GC likewise passes risk, including schedule risk, to its subcontractors through contract language. To be successful, the project manager and the superintendent must identify the potential project risks and select strategies for managing them. The potential consequences of these project risks may pose threats to project success, but using a proactive approach to identifying the risks and potential consequences of their occurrence enables selection of strategies for mitigating the effect on project success.

Construction is a risky business for a variety of reasons and is proven out by the high number of construction firm failures each year. To minimize the potential for financial difficulty, a contractor should analyze each potential project to determine the risks involved and whether or not the potential rewards justify acceptance of the risk exposure. Risk management involves risk identification, measurement, and mitigation strategies.

The first level of risk analysis occurs when the contractor decides whether or not to pursue the project. While some business risks can be transferred by the purchase of insurance coverage, most project risks are not transferable. Many GCs transfer the risks associated with specific scopes of work to subcontractors; however, the GC retains responsibility for the timeliness and quality of the subcontracted work. Selection and management of subcontractors will be discussed in Chapter 17. Risks typically encountered during construction are classified as either external or internal risks.

The sources of *external risk* on a project may involve things like unusually adverse weather, material cost inflation, owner's inability to finance the project, limited availability of skilled craftspeople or subcontractors, performance and bankruptcy of subcontractors, incomplete design documents, project location, theft and vandalism, safety, environmental requirements, and project complexity. The contractor needs to forecast the likelihood of such risks, the range of possibilities, and the impact of each on the contractor's ability to complete the project profitably. Projects with increased risk require a larger fee in return for the risk accepted than those with less risk. Methods to mitigate external risks include:

- Employing subcontractors,
- Purchasing insurance,
- Purchasing performance and payment bonds for subcontractors, and others.

Internal risks will also be identified and must be managed both at the corporate and project levels. The most common internal risks are unrealistic cost estimates, unrealistic construction schedules, ineffective jobsite management, crew performance, quality control, equipment operation, and schedule adherence. Jobsite management shortcomings also include cost and schedule control, material management, and subcontractor coordination.

Contractors must adopt strategies to minimize the potential of these problems occurring. Often the basic issue to be addressed is the selection of qualified people to manage the project, particularly the PM and superintendent. The output of a risk analysis process results in a decision whether or not to pursue a project; the amount of fee and contingency to include in the bid or cost proposal; additional contingency time built into the schedule; whether or not to joint-venture with another firm; the portions of work to subcontract; and the type and amount of insurance to purchase. If the contractor cannot meet the owner's contract schedule requirements they may choose to pass on the job – simply shortening activities or squeezing the subcontractors or waiting for the owner and/or architect to impose delays is too risky.

6.8 SUMMARY

The contract documents have a significant impact on the responsibilities of the project management team. The prime contract is the agreement between the GC or CM and the project owner. It is a legal document that describes the rights and responsibilities of the parties. The majority of the focus in this chapter has been on the prime agreement, but many of the documents and processes described apply to subcontract agreements and material supplier purchase orders as well. All contract provisions are important and the estimator and scheduler need to have their corporate officers review the contract in detail before submitting a bid or signing a contract – often with legal assistance.

Construction is a risky business, and contractors must carefully assess the risks associated with each prospective project. Once the risks have been identified, risk management strategies must be developed. In some cases, the risks are too great, and the project should not be pursued. In other cases, the risk can be mitigated by obtaining a joint-venture partner or hiring subcontractors or with increased fees.

This concludes the final chapter in the planning part of this book. After all of the initial preconstruction and planning is complete, including addressing lean considerations and contract issues, the scheduler will now prepare the schedule document. Part III expands on different types of schedules and the process to develop a schedule, including all of the calculations that determine the critical path and potential float. The scheduler must balance schedule needs with available resources, which includes cash flow. Today most schedules are prepared using scheduling software and these methods are also discussed in the next part of the book.

6.9 REVIEW QUESTIONS

1. Which family of contracts favors the architect and owner more so than the contractor, and which favors the contractor more so than the architect and owner?
2. Which agencies/firms issue a TCO, C of O, and/or certificate of substantial completion?
3. How does a project owner pass risk to a GC? How does a GC pass risk to a subcontractor?
4. Why would a GC choose to insert a LDs clause into their contract?
5. What is the difference between a work day and a calendar day?
6. Day one of a schedule should be predicated on what four items?
7. What are the five requirements for a contract to exist?
8. What are the five primary contract documents?
9. How can any document be added to a contract, including a cocktail napkin, (A) pre-bid, and (B) post bid?
10. Which contract document will tell the GC its work day and hour limits?
11. What day or type of day is not a "work" day? There are several potential answers.

6.10 EXERCISES

1. On a project with liquidated damages and bonus clauses, if additional days were awarded the GC due to conditions beyond their control (e.g. weather or changed conditions), say plus five days was awarded the GC but the GC still finishes ahead of the original contracted schedule finish date, are they awarded a bonus? If additional jobsite general conditions had also been awarded, but were not expended, would the GC owe those dollars back to the project owner? Do your answers change if this was a negotiated versus lump sum project?
2. Why might a city shy away from issuing a TCO?
3. Why does ConsensusDocs 751 appear to take a collaborative approach to subcontractors with respect to scheduling?
4. Why is it better to use a standard copyrighted contract format than a home grown contract? Conversely, why might it be advantageous for large procurers of construction services to develop their own contracts?
5. If your lump sum project had a liquidated damages clause, what actions might you take over the course of construction to protect your firm?
6. Who owns the float?

Part III

Scheduling

Part III

Scheduling

Chapter 7

Schedule Types

7.1 INTRODUCTION

The last group of chapters covered many different types of schedule planning processes undertaken during the preconstruction period, including planning for implementation of lean processes and evaluating proposed contract language for impacts on scheduling. The result of planning is a list of construction and nonconstruction activities arranged in a logical order. Now this group of chapters progresses with drafting the schedule by adding durations and time lines, calculating the critical path, incorporating resource limitations, and includes the effect the schedule has on cash flow projections. This chapter discusses many different schedule formats, their uses, and advantages and disadvantages as construction communication tools. Chapter 1 briefly introduces the reader to a variety of different types of schedules, many of which are discussed throughout this book. There is not one perfect schedule for all situations. It is not a "one size fits all" process.

There are two primary types of schedules: bar charts and network diagrams. Bar charts relate activities to a calendar, but often show little to no relationship among the activities. Network diagrams show the relationship among the activities, and may or may not be time-scaled on a calendar. Two techniques are used in developing network schedules. The first is known as the arrow diagramming method (ADM) in which arrows depict the individual activities. The other is known as the precedence diagramming method (PDM) in which the activities are represented by nodes. The arrow diagramming method used to be prepared manually, but today both the arrow and the precedence diagramming methods are computer-generated schedules. Network diagram schedules are sometimes referred to as "critical path method" (CPM) schedules. The critical path is the longest path through the schedule and determines the overall project duration. Any delay in an activity on the critical path results in a delay in the completion of the project. Critical path scheduling began in the late 1950s. The Navy developed PDM and production industries utilized ADM. All of these methods were discovered and improved by industries, such as car manufacturing, and governmental agencies, more so than the construction industry. The next three sections of this chapter explain the process to develop bar charts, ADM, and PDM network schedules.

The type of schedule format and system utilized by the jobsite management team is influenced by the construction company's requirements and resources, such as available software. The type and size of the construction project and the individual superintendent's scheduling skills will also play a part. In addition, some project owners will require a particular schedule type. For example, a contractor building an office building in the Pacific Northwest for Microsoft (MS) will likely use MS Project scheduling software. The client in many public works projects determines many of the schedule requirements including start and completion dates, quantities of activities, maximum duration of activities, frequency of updates and software requirements. The fire station case study's special conditions to the contract required the successful general contractor (GC) to use MS Project scheduling software. Other construction consumers, such as the United States Army Corp of Engineers (USACE), require contractors to use Primavera by Oracle.

Schedules can take on a variety of formats, including a simple tabular list of activities and dates, but these are sometimes difficult for project participants to understand. A tabular excerpt from the case study project detailed schedule is included as Figure 7.1. Most builders expect a bar chart or network diagram that graphically depicts work starting from the left and proceeding to the right, and starting from the top of the page and proceeding toward the bottom. The net effect will be a "waterfall" of activities connected by restraint lines.

Mountain Construction Company
Tabular schedule

Project: <u>Fire Station 83</u>   Date: <u>8/12/22</u>
Superintendent: <u>Ralph Hendrix</u>   Sheet: <u>1 OF 1</u>

No.	Activity description:	Start date	Work days Duration	Finish date	Comments
10	SOG finegrade	8/12	3	8/16	Prioritize 8" apparatus
15	Vapor barrier under SOG	8/17	2	8/18	SOG first
20	Underslab utilities	8/18	3	8/22	Plumber, Electrician
25	Formwork	8/18	4	8/23	
30	Reinforcement steel	8/23	3	8/26	
35	Place/Pump/Finish 8" SOG	8/26	1	8/26	
40	Place/Pump/Finish 4" SOG	8/29	1	8/29	
45	Spray curing compound	8/29	2	8/30	
50	Remove forms	8/31	1	8/31	
55	Backfill at edge	9/1	1	9/1	
60	Polish SOG	9/2	5	9/9	Mobilize sub prior

Figure 7.1 Tabular schedule

A summary schedule is often included with the project manager's (PM's) monthly internal report to company executives. This monthly report will also include the budget analysis and fee projection, updates on safety and quality controls, relationships with the project owner, change order and pay request management, and status other potential areas of risk management. The summary schedule is also a communication tool between the GC and the project owner to status the project on an executive level.

The best format for a schedule depends upon its use as a communication and construction tool. The most appropriate schedule format considers the audience and the information necessary to communicate. Webster loosely defines a schedule as a written document that incorporates details and a time line. All schedule formats are good systems and may be appropriate in different applications. Some of the schedule formats discussed in this chapter include:

- Contract schedules,
- Formal schedules,
- Summary schedules,
- Detailed schedules,
- Short-interval schedules,
- Pull schedules,
- Specialized schedules,
- Mini-schedules,
- As-built schedules, and others.

7.2 BAR CHARTS

The bar chart is also known as a Gantt chart and was named after Henry Gantt who created it in 1917. The bar chart can be presented in other formats including graphs and curves and is often utilized for resource leveling. The major advantage of the bar chart is its simplicity and clear means to communicate the plan and the schedule to those who need the information, including those unaccustomed to reading schedules.

An early bar chart was a graphical extension of the logic diagram prepared during preconstruction, as discussed in Chapter 4. The scheduler must get the logic relationships correct first. By simply adding a time line to the logic diagram, and approximated activity durations, the scheduler has created an early rough draft of the schedule. It is best to begin with a pure logic diagram without dummy activities. Bar charts are typically thought to have limitations, such as they lack relationships. But relationships, or vertical restraint lines, can be added; they just must be coordinated such that they do not dominate the network.

Figure 7.2 reflects the same logical activities in a bar chart but with three different approaches. The first represents the slab-on-grade (SOG) activities in a chain format, each predecessor linked to its successor. The second is a "stair step" or "waterfall" format that facilitates listing activity numbers and descriptions along the left-hand side of the drawing. The third format adds vertical restraint lines, or relationship lines to the second.

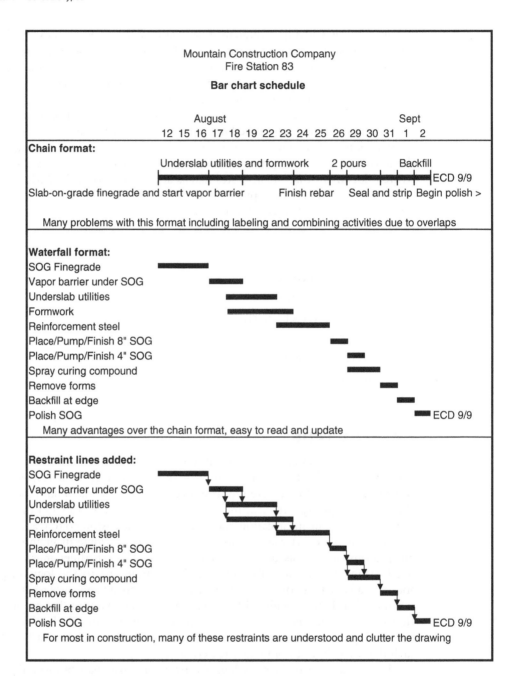

Figure 7.2 Bar chart schedule

A summary bar chart is a useful tool for corporate executives when displaying multiple projects on the same schedule, but not cluttering it up with detail. This allows evaluation of current and future backlog or workload. It also is an ideal planning tool for the contractor's most valuable resources, their people, and specifically their superintendents and project managers. Each single project schedule line is backed up by a summary schedule, which refers to another detailed schedule that is posted to each project's jobsite trailer conference room. Each PM customarily prepares monthly reports to his or her management, which includes a fee forecast and an updated schedule, among other documents. The home office can then incorporate this into their corporate planning tool. An example of this corporate resource schedule is included in Chapter 10.

7.3 ARROW DIAGRAMMING METHOD

An early critical path schedule was displayed through the use of the arrow diagramming method, also known as the activity on the arrow (AOA) method. The ADM is similar to a bar chart with restraint lines (option 3 in Figure 7.2) but adds nodes at the start and end of each activity and arrows at the end of the bar or activity. Nodes depict events. An activity starts at one end of the arrow, typically the left end, and this is an event, and completes at the other end of the bar, the end with the arrow typically on the right side, and this is the other event. The duration of the activity is represented by the length of the horizontal bar; a five-day activity has a longer bar than a two-day activity.

In lieu of activity numbers, such as activity #15, the ADM method utilizes numbers from each of the predecessor and successor nodes. These are referred to as the "ij" numbers or nodes. It is customary that round numbers with plenty of space between are used, not 22, 23, 24, but rather 10, 20, 30, which allows for the introduction of additional activities at a later point of schedule development (e.g., a missed restraint) or relationship, such as a material delivery. The activity number then takes on the combination of the two ij node numbers such that placing the rebar in the SOG is activity 25-30. Because of the ij nodes, the ADM may be also be referred to as the ij scheduling method.

No two activities can have the same set of ij numbers. Each combination must be unique. If two activities have the exact same predecessor and successor nodes, then a *dummy activity* is introduced. A dummy activity is not a construction activity but represents a relationship. The dummy activity itself does not incur time and does not utilize resources. A criticism of the ADM scheduling method is the need to add these dummy activities which may increase the activity count by up to 15%.

The ADM scheduling approach typically has only one activity the network starts with and only one activity the schedule completes with. Therefore, there is just one start node, often number 10, and one end node (event) for the entire network.

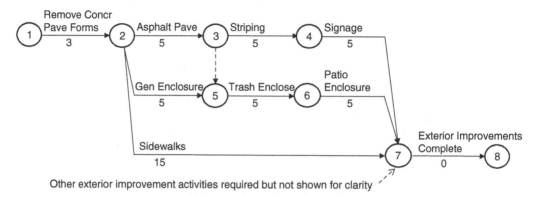

Figure 7.3 Arrow diagramming method example

Schedulers who utilize the activity on the arrow method must truly understand construction relationships and the necessary use of dummy activities to correct the schedule logic. All of this can be complicated and time-consuming but not insurmountable if the scheduler also has field experience. An interesting side note is that some scheduling authors say ADM is antiquated and not used in construction, although ADM is used by instructors to easily explain and teach scheduling concepts, especially relationships and time relevancy. A couple of full-time schedulers queried when researching this book also indicated it may not be necessary to discuss ADM as it is no longer used. But bar charts with restraint lines connecting activities is similar and relevant. Arrow diagramming method, without the use of activity start and end nodes, is essentially a time-scaled bar chart with restraint lines. This hybrid schedule format is very common on construction projects. Arrow diagrams, or AOD schedules, in some format therefore are still utilized on the construction site. This book also utilizes several ADM sketches in Chapter 9 to discuss schedule calculations, including critical path and float. An example of an ADM network diagram is included as Figure 7.3. A brief understanding of scheduling history is a good platform for many of the more popular scheduling tools that will be discussed in this book.

7.4 PRECEDENCE DIAGRAMMING METHOD

The precedence diagramming method was also developed in the 1950s. This network format is also known as activity on the node (AON). Precedence diagrams do not utilize start and end nodes, as is the case with ADM, but rather the entire activity is one large node. The ADM and bar chart activity bars are replaced with an activity node, which can be a large rectangle or a circle.

The activity on the node display resolves the need for additional dummy activities associated with the activity on the arrow method. Because of the lack of the need for dummy activities, some schedulers find PDM easier to diagram, read, and manage than ADM. Each activity is completely represented by one single larger node in lieu of a bar or arrow.

Activity on the node is a network without a time frame. Precedence diagrams do not necessarily have a graphical and scaled time line as does ADM. Contractors think in terms of a time line, which typically moves from left to right. Contractors are also accustomed to a bar chart or bar chart with relationships similar to an ADM.

Precedence diagrams are easy to draw, especially during the planning stage and logic development, but although relationships can be incorporated into this chart, they can be complicated. The advantage of a PDM is the amount of information that can be included on the node with respect to activity descriptions, duration, early and late start and finish dates, total and free float, and others. But the double-edged sword analogy of all that information is that the schedule can become overwhelming. The construction schedule must remain a useful tool to assist contractors with building a project. An example of a PDM draft network is included in Figure 7.4.

Some scheduling enthusiasts and authors indicate that PDM is "sufficient for a scheduler to manage a project." This author takes exception with that statement as schedulers do not manage construction projects – builders do. Although most

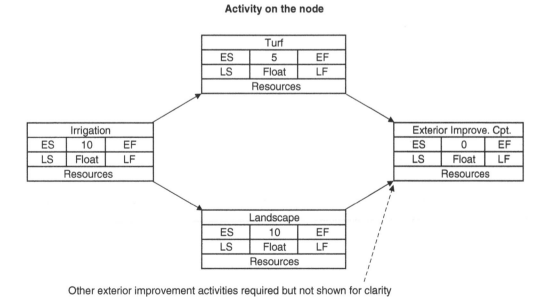

Figure 7.4 Precedence diagramming method example

other scheduling textbooks indicate schedules today are PDM and not ADM, this author, professor, and construction industry professional for over 40 years has never seen a PDM schedule on a jobsite. Today most construction schedules are bar charts with the addition of vertical restraint lines. Many professional schedulers, university professors, and academic authors have their preference as to which is the most "correct" scheduling method: bar chart, ADM, or PDM. Table 7.1 summarizes a few of the differences between the three systems and reported advantages and disadvantages for each. All three will be utilized in further examples in this book.

Table 7.1 Schedule type comparisons

Conventional scheduling discussions compare the major schedule types and processes along the following lines:

Bar chart advantages over networks:
- Time scaled
- Easy to prepare
- Easy to understand
- Presentations
- Load with resources

Network advantages over bar chart:
- Logic
- Larger projects
- Dates based on CPM
- Show relationships between activities

Advantages of activity on the arrow networks over node networks:
- Definable activities and events (start and finish)
- Milestones are easy to show
- Time-scaled

Node networks, or precedence diagrams, have the following advantages over arrow networks:
- Easier to read and draw
- No need for dummy activities
- Activity ID
- Neater logic
- Includes resources
- Easier to show lag between activities
- Easy to show relationships, such as start to start (SS) or finish to finish (FF).

7.5 CONTRACT SCHEDULES

Formal schedules may be developed and submitted to the project owner as required by the contract or prepared and submitted with a proposal for a negotiated contract. The official *contract schedule* is typically detailed, but some contracts, such as a negotiated contract, may require only a *summary schedule*. Some contractors would prefer providing only the project owner and designer with a summary schedule, if allowed. Even if the schedule is detailed, the contractor should still practice the 80-20 rule, such that 80% of the construction time and focus will be attributed to 20% of the project scope. It is that 20% that should be reflected in the detailed contract schedule. Forming and pouring the concrete foundation walls should definitely be shown on the contract schedule, but spraying form oil on the forms before a pour is not necessary.

Some contracts will prescribe the schedule format, required scheduling software, quantity of activities, and method and timing for updates and revisions. As discussed in the previous chapter, the schedule document should be referred to and attached to the prime contract as an exhibit and likewise should be included in each subcontract agreement. This reference is likely more common on private than public projects. But similar to a contract drawing, if the schedule is modified, the prime agreement and all subcontract agreements should be modified, which opens the door for potential change orders and claims. The contract schedule is also known as the baseline or record schedule. There are many different types of record schedules just as there are many different types of record drawings, such as bid drawings, permit drawings, submittal drawings, as-built drawings, and others. It is unfortunate, but one purpose for a contractor producing an excessively detailed schedule is to set itself up for preparing a postproject claim and to prove schedule impact and extension. Summary schedules, like the one illustrated in Figure 7.5 often are used for presentations, management reporting, or as an exhibit to a contract. This is an expanded summary schedule version compared to the case study's milestone schedule included with Chapter 1.

The contract schedule should be a tool utilized by all of the team members and should be hung on the GC's jobsite trailer wall and project status shown and discussed at the weekly owner-architect-contractor (OAC) meeting by the project superintendent. One problem with a schedule that is too summarized is that it is difficult to prove schedule impact from scope changes or delays in decisions from the owner or design team. A *detailed schedule* should be the version posted on the wall of the meeting room or in the jobsite trailer. They are marked up with comments and progress. The balance between a detailed and summary schedule is reflected in the next example. The schedule should be looked upon and utilized as a communication tool to help build the project. The detailed schedule for the case study project is included in the book's website.

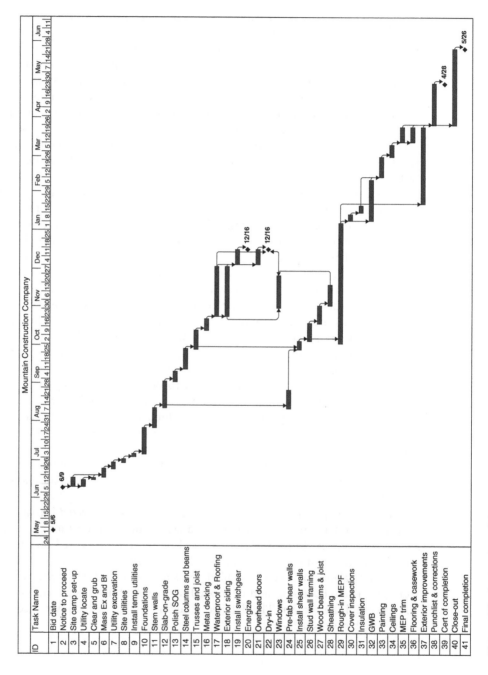

Figure 7.5 Contract exhibit summary schedule

Example 7.1

It is important that the project or contract schedule be looked upon as an overall guideline but it does not typically represent the day-to-day planning that builders rely on. A young superintendent was on a high-profile project and would be visited twice-weekly from the company's chief operating officer. The GC executive would point to the schedule on the wall and grill the superintendent why he was one day late on an activity or two and did not address others that might have even been a week ahead. The superintendent would explain, and was defensive, but would later in his career avoid putting any detail into a contract schedule for fear that he would later be challenged, not only by a client, but by his own executives.

7.6 SHORT-INTERVAL SCHEDULES

Short-interval or *look-ahead* schedules are developed weekly by each superintendent or foremen and each subcontractor. They may be hand-drafted in bar chart form, electronically through Microsoft Excel, or with scheduling software like MS Project or Primavera Project Planner. It is critical that short-interval schedules are produced by those who are performing the work and are communicated and distributed to the project team. In this way, short-interval schedules can be effective construction management communications tools. The format and/or software used to develop the short-interval schedule is of less importance than the author. The timescale of these types of schedules can be two-three-or four-week increments, depending upon project requirements and level of activity. Figure 7.6 shows an example of a three-week SOG schedule from the case study project. The superintendent for this project prepared this schedule utilizing Excel.

Construction is not really a "production" industry, but some refer to three-week look-ahead schedules as short-interval production schedules (SIPS). SIPS is also an abbreviation for structurally insulated panel system and in some jurisdictions, street improvement permits. This book will stay away from SIPS and refer to this scheduling tool as a short-interval or three-week or look-ahead or foreman's schedule.

While contract schedules, such as those previously discussed, are developed for overall project control, short-interval schedules are used by the superintendent to manage the day-to-day activities of the project. These schedules are developed each week throughout the duration of the project and typically provide sufficient information for managing the project from a field supervisor's perspective. Neither the title, nor the format are as critical as who prepares the schedule. The preparation of short-interval schedules is the responsibility of the last planners: The foreman and super-intendents who are responsible for the construction work. It is imperative that they have buy-in with the schedule. The schedules should not be delegated to the project engineer, but the project engineer may assist as necessary.

Mountain Construction Co.
Short-interval schedule

Project:	Fire Station 83, Pasco, WA	Project number:	9922	Date:	8/15/22
Superintendent:	Ralph Henry			Sheet:	1 of 1

No.	Activity description:	S	S	15	16	17	18	19	S	S	22	23	24	25	26	S	S	29	30	31	1	2	S	S	Comments
				August, 2022														Sept, 2022 >							
1	Begin SOG finegrade Fri, 8/12	X	X						X	X						X	X						X	X	Prioritize 8" apparatus
2	Continue finegrade	X	X						X	X						X	X						X	X	bay SOG first!
3	Vapor barrier under SOG	X	X						X	X						X	X						X	X	
4	Underslab utilities	X	X						X	X						X	X						X	X	Plumber, Electrician
5	Formwork	X	X						X	X						X	X						X	X	
6	Reinforcement steel	X	X						X	X						X	X						X	X	
7	Place/Pump/Finish 8" SOG	X	X						X	X					8"	X	X						X	X	
8	Place/Pump/Finish 4" SOG	X	X						X	X						X	X	4"					X	X	
9	Spray curing compound	X	X						X	X						X	X	8"	4"				X	X	
10	Remove forms	X	X						X	X						X	X						X	X	
11	Backfill at edge	X	X						X	X						X	X						X	X	
12	Polish SOG	X	X						X	X			Mobilize sub prior:										X	X	ECD 9-Sept

Figure 7.6   Short-interval schedule

Subcontractors should be required to prepare and submit to the project team their three-week schedules at the Monday morning foremen's meeting. The GC's project superintendent then collects all of the subcontractor schedules and summarizes and presents his or her three-week schedule at the weekly OAC meeting.

One unfortunate method utilized by some PMs to respond to the requirement to submit a weekly look-ahead schedule to the project owner is to simply take a three-week window from the detailed computer schedule without adding any additional updates or detail, and hand that out at the weekly owner-architect-contractor meeting. An example of this is included as a figure in Chapter 13.

Pull schedules, prepared by the last planners, are lean construction tools that have been adapted from the automobile production industry. This type of short-interval schedule was discussed in Chapter 5. To-do lists are also great short-term scheduling tools. To-do lists should not be overly full or long as each jobsite manager must have room and flexibility to deal with today's unknowns or "fire drills." For some professionals, a very simple alternate short-term schedule format is the use of a calendar. Several of these short-interval scheduling tools will be discussed again in Chapter 13 with schedule controls.

7.7 SPECIALTY SCHEDULES

Specialty schedules are mini-schedules that focus on a subset of the entire project and include: Area schedules, system schedules, subcontract schedules, start-up and commissioning schedules, close-out schedules, and others. Specialty schedules allow additional detail for certain portions of the work that could not be adequately represented in the project schedule and have longer durations than the short-interval schedule. Specialty schedules are a cross between the contract schedule and three-week look-ahead schedules. One schedule control method is to expand and contract portions of the contract schedule as the project progresses, e.g. during precover mechanical, electrical, and plumbing rough-in, inspections, and insulation. Activities can be added to provide more detail to a schedule to better reflect the requirements of the work at hand. A sample specialty schedule for the case study project's roofing scope has been included on the book's website. Examples of specialty schedules include:

- Milestone schedule (example included with Chapter 1),
- Preconstruction schedule (Chapter 3),
- Schedule of project phases,
- Schedule for building wings,
- Schedule for building floors,
- Separate the site schedule from the building schedule,
- Separate shell and core schedule from tenant improvement schedule,
- Construction startup or mobilization schedule,
- Submittal schedule or submittal log (Chapter 14),
- Deferred submittal and specialty permit schedules,
- Expediting, procurement, and/or buyout schedules, also referred to as logs (Chapter 17)
- Shutdown schedules, popular with civil and industrial projects,
- Mechanical, electrical, and plumbing equipment start-up schedules,
- Elevator schedules,
- Heating, ventilation, and air-conditioning balancing and commissioning schedules,
- Owner furnished materials and equipment schedule: Deliveries, installations, and start-up, and
- Close-out schedule or log.

Subcontractors should be responsible to contribute their own schedules, rather than the general contractor's project superintendent dictating his or her requirements to the subcontractors. This is one form of collaborative versus top-down scheduling and CM process. If time allows a GC on a negotiated project to receive and incorporate subcontractor input before finalizing the contract schedule to the project owner, the project superintendent will achieve subcontractor buy-in.

A *fragnet* is a partial schedule, or portion of a larger network schedule. Fragnets may be used for paths or sequences that repeat themselves. An example would be a 200-home subdivision that builds 10 different plans. There would likely be a detailed schedule for each of the 10 different designs, but not 200 detailed schedules. And the overall project schedule would have just one-line item for each home and refer to the separate fragnets for details. The term *fragnet* is a scheduling term, essentially meaning a fraction or fragment of a larger schedule or network. Fragnets are also known as subnet, subnetwork, subcontractor schedules, specialty schedules, or mini-schedules. Some government agencies, such as the USACE, require submission of fragnet schedules to accompany any requests for additional time and/or compensation.

A *rolling schedule* is another version of a specialty schedule. Contract schedules are often a summary schedule and can refer to other detailed subnet schedules. A contract summary schedule is essentially a rolled-up schedule where each line item can be expanded either with reference to another detailed schedule or expanded when that portion of the work is in the near future. The scheduler cannot include all administrative or nonconstruction jobsite work activities in one complete detailed schedule especially on larger projects. The schedule would be too detailed, including activities outside of the 80-20 rule and therefore not a useful communication tool for foremen and subcontractors. These other essential activities can be included on the computer, but temporarily hidden for presentation purposes. The solution is to be able to sort correctly, and expand certain areas with additional details that are normally hidden, summarized, or rolled up. An example of this is shown in Figure 7.7, which blows up the fire station shell construction activities in greater detail than shown in the previous site work or the following finishes. Other portions of this schedule can be blown up even further and are included in other three-week or fragnet schedule examples and figures throughout this book.

Another specialty schedule that is not always developed is an *as-built schedule*. This schedule is ideally created throughout the course of the construction project by recording actual delivery dates, actual start and completion dates, milestones realized, and potentially subcontractor manpower. This schedule is an excellent element to include with the post-project lessons-learned report. The as-built schedule may also be utilized in claim preparation or defense – so it is important that only verifiable data be recorded. As-built schedule development and claims are discussed in Chapter 18. Many of these other specialized schedules are also discussed in other chapters in this book.

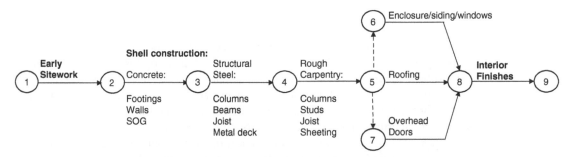

Other shell and enclosure activities required but not shown for clarity

Figure 7.7 Rolling schedule

7.8   SCHEDULE FORMAT

A project plan can be developed in many different formats. It can be an outline, notes, sketches, scribbles on butcher paper, meeting minutes, or a roughed-out schedule, either with or without a time line. This information is then provided to the scheduler to input into the selected scheduling software program. In some instances the scheduler may be the same individual as the planner. On some projects, the PM, in conjunction with the superintendent, may develop the schedule. The duration for each activity is determined using the crew productivity factors from the cost estimate. Now the start and finish dates for each activity can be established, and the overall scheduled project completion date is compared with the contractual completion date. Also, all the activities that cannot be completed late without delaying the project completion can be identified. These activities will be laid down along one or multiple sequential paths from start to finish, which is called the critical path. If the schedule shows completion later than the contract requires, some of the activities on the critical path must be accelerated or stacked to produce a schedule that meets contract requirements.

Very few schedules are hand-produced anymore, although early drafts developed in the planning process may still be done this way. But as reflected in the next example, the schedule format is not as important as the knowledge and skills of the planner and scheduler. Today, most schedulers use computers to produce the final schedule product. The identification of the activities, the duration for each, and the sequence in which they are to be completed must be input into the computer for it to properly produce the schedule. The computer does not plan the project, but it will plot the schedule and calculate the start and finish time for each activity. Computer-generated schedules allow the scheduler, PM, and superintendent to quickly determine the effects of changing schedule logic, delays in delivery of critical materials, or adjustments in resource requirements. The computer is a valuable scheduling tool, but it cannot take the place of adequate planning.

Example 7.2

A small sole-proprietor general contractor in Alaska was awarded a large commercial office building on a negotiated basis due to his wonderful reputation in the community and personal connections with the owner and architect. He was both the project manager and superintendent on all of his projects and also managed a very busy panelized prefabrication facility. The contracted owner's representative was from Seattle, Washington, and expected project controls sophistication comparable with his Seattle GCs. The Juneau contractor had never produced a three-week schedule and resisted the "big-city" oversight he anticipated receiving. The owner's representative helped him through this and eventually received buy-in to use this valuable tool. The project was a huge success on many fronts. His schedule was not pretty – it was drawn by hand- scrawled over with phone numbers and marked by coffee spills, but still it was his. On a subsequent project proposal, he included a sample of his three-week schedule as one of "his" standard control tools; and he landed that project as well!

7.9 SUMMARY

Schedules can take on many different formats and serve different purposes. Sometimes this is predicated by the contractor's culture, contract requirements, or likely what each scheduler is comfortable using. A bar chart schedule was one of the first construction schedule formats and is still in use today. It remains a viable construction communication tool. The addition of vertical restraint lines adds to its complexity and resolves some of its shortcomings.

The arrow diagramming method is time-scaled and is similar to a bar chart except the arrow (which represents activity durations) is preceded and followed by activity nodes, or ij nodes. These nodes represent event times – the start and completion of each activity. The ADM often requires introduction of dummy activities which do not reflect work but rather relationships. The PDM schedule format places all of the information about an activity in one much larger node than the ADM. Nodes can be circular or rectangular and will include activity numbers and/or descriptions, durations, start and finish dates, float calculations, resources, and others. Precedence diagrams are not normally time-scaled and do not require the introduction of dummy activities. All three have advantages and disadvantages and may be adaptable to different projects, contractors, and schedulers in a variety of fashions.

The most current contract schedule for negotiated and bid projects should be the detailed schedule that should be displayed on the wall of the jobsite conference room and statused at the weekly OAC meeting. A summary schedule is often required of negotiated requests for proposal by the project owner and typically included in executive progress reports. The original bid schedule, contract schedule, and as-built schedule are all record documents and should be kept secure for future reference. The schedule document will be incorporated into the prime contract agreement and each subcontract and purchase order, just as drawings and specifications are incorporated into contracts. For public projects the baseline contract schedule is the one initially submitted and approved by the owner. Change orders may result in revisions to the contract schedule. In addition to creation of the original schedule, the superintendent will status, update, and revise the contract schedule during the course of construction.

Three-week look-ahead schedules are developed by foremen and subcontractors and used to communicate daily activities including deliveries. The GC's project superintendent presents his or her short-interval schedule at the weekly OAC meeting. This detailed schedule is a compilation of all of the subcontractor three-week schedules that were exchanged and discussed at the weekly Monday-morning foremen's meeting. There are numerous other specialty scheduling tools available to jobsite management teams including area schedules and expediting logs. These schedules are also more detailed than the contract schedule.

Although the schedule format is not as important as the message and the messenger, the presentation needs to be clear and concise enough to prove a useful communication tool. The schedule is a tool to communicate the original plan, report on the current status of the project, and forecast the path to its successful completion. The scheduler must know his or her audience. The scheduler must know their crowd, whether that be the city, subcontractors, architect, foremen, GC's home office, project owner, and others. The next chapter takes the result of planning and progresses the logic diagram into an early schedule draft.

7.10 REVIEW QUESTIONS

1. Explain the 80-20 rule with respect to schedules.
2. What does "critical path" mean?
3. Who is the best person to create the schedule plan and why him or her?
4. Who should prepare a three-week look-ahead schedule for the plumbers' work, and why him or her?
5. Why should the project superintendent, and not the project manager or home office staff scheduler, status the schedule at the weekly owner-architect-contractor meeting?
6. What is the difference between a three-week look-ahead schedule and a pull planning schedule?
7. Match these schedule abbreviations up into two pairs: ADM, AON, AOA, and PDM.
8. How is a fragnet different from a short-interval schedule? They have many similarities.

7.11 EXERCISES

1. What does the term *play to the crowd* mean as it relates to scheduling?
2. This chapter somewhat contradicted itself. Is the schedule format important or not?
3. Why did the case study superintendent include non-roofing activities on the specialty schedule included on the book's companion website?
4. In addition to schedule, what other aspects of the project might the GC's superintendent report on at the weekly OAC meeting?
5. How many activities should a contract schedule have?
6. Draw three different logic diagrams utilizing a simple bar chart method (without restraints), ADM, and PDM for the following concrete retaining wall work activities, which are listed in logical order: layout wall, form one side, place rebar, form the second side, brace the forms, place concrete, set anchor bolts, rod-off concrete, cure, strip forms, remove snap ties, patch and sack as necessary. Assume a standard two-day activity duration for presentation purposes.
7. Compare the three-week schedule in this chapter with that from Chapter 1. Additional versions will be presented in Chapter 13. Does one format or another appeal to you better? What other formats of three-week schedules have you witnessed out on the jobsite?

8. A live version of summary schedule Figure 7.5 has been provided on the companion website. Can you propose a rearrangement of activities that (A) further minimizes crossing of horizontal activity lines with vertical restraint lines, and/or (B) presents a more efficient arrangement of start and/or finish dates?

9. Other than roofing, mechanical, or electrical, what areas or subcontractors from the case study project warrant a specialty schedule?

10. Many contractors simply utilize a three-week software "sort" from the detailed schedule to present at the OAC meeting. Why might this not be the best construction communication tool?

11. Prepare a to-do list for your day tomorrow, commencing with your alarm ringing and concluding with setting your alarm in the evening. Don't leave anything out. Check each item off as the day progresses. Feel free to add new items as they come up. Compare with your classmates. Who had the most action items?

Chapter **8**

Schedule Development Process

8.1 INTRODUCTION

Time management is just as important to project success as is cost management. The estimate and schedule are interrelated on many levels. The key to effective time management is to carefully plan the work to be performed, develop a realistic construction schedule, and then manage the performance of the work. This chapter discusses schedule preparation and Chapter 13 covers schedule control. Chapter 9, "Schedule Calculations," is separated from this chapter but is integral with the schedule development process.

The schedule is worthless as a construction tool unless it can be clearly communicated; it must be neat, logical, and clean. The contractor should not hide the schedule from any of the internal or external team members for fear that missing a date or two here or there will invite criticism. There should be only one detailed schedule that all parties can trust and rely upon. There should not be "two sets of books," that is, an internal and a separate external schedule. The schedule is a communication tool that conveys the plan to many of the project stakeholders as indicated here:

- The schedule is a tool to inform the project owner when decisions are necessary.
- The design team needs to be apprised of project timelines when processing submittals and responding to requests for information.
- The city appreciates advanced scheduling to plan for inspections.
- Subcontractors utilize the schedule to anticipate when materials should be delivered and craftsmen mobilized.
- The project owner needs the schedule for delivery of fixtures, furniture and equipment (FF&E), advise tenants of move-in projections, and arrange permanent financing.
- The project manager (PM) will use the schedule and the estimate together to produce a cost-loaded schedule and cash-flow curve for the project owner and the bank to make sure sufficient construction funds are available to support monthly pay requests. Cash flow is the topic of upcoming Chapter 10.

Example 8.1

The lead scheduler was only 25 years old on this $3 billion power plant pro-ject. He had 17 schedulers working for him. An older and more experienced scheduler was also on his team and did not appreciate reporting to a younger person. He resisted the younger boss's attempts to review his schedules before presenting them to other stakeholders – he felt he knew everything there was to know. So the lead scheduler decided to let him be. During one of his presentation meetings the superintendents criticized the older scheduler immensely. The scheduler from then on ran drafts of his schedules past his boss and invited him to future presentation meetings.

There are many different formats a schedule can take, and any one project may have several different schedule types. Many schedule methods and format were discussed in the last chapter. The prime contract agreement should refer to an estimated cost and a schedule duration or completion date. But both the estimate and the schedule documents should also be included as contract documents, similar to drawings and specifications. Scheduling is the act of preparing the schedule document. The schedule development process involves collecting all of the planning information and plotting the results on paper or on the computer. Also similar to preparing a cost estimate, schedules should experience many draft copies and reviews before being finalized and shared with other built environment stakeholders. No one is above the need to prepare drafts of their work and receive critique or input, as Example 8.1 indicates. The ability of the jobsite team to produce a good construction schedule, which functions as a key communication tool, is an essential construction management skill.

8.2 SCHEDULE PLANNING

Many see planning and scheduling as the same operation, but they are slightly different with proper planning preceding development of the project schedule. Planning is usually performed by the project manager and superintendent with the assistance of other experienced and specialized contractor personnel, such as a staff scheduler. Planning includes performing the following tasks:

- Developing a work breakdown structure (WBS);
- Identifying a logical flow of work activities and incorporating restraints;
- Evaluating the availability of manpower for self-performed work;
- Developing a subcontracting plan, including what work is to be subcontracted versus self-performed;
- Estimating activity durations from labor hours off the direct-work pricing reca-pitulation sheets and subcontractor input;

- Forecasting material and equipment delivery dates;
- Identifying owner and architect restraints, such as design package releases, receipt of permits, and delivery of owner-supplied FF&E; and
- Selecting means and methods of construction including choices of concrete formwork, internally or externally rented equipment, and hoisting (tower crane, crawler, boom truck, and/or forklift).

The development of the actual schedule document is a simpler and more mechanical process than developing a proper project plan as reflected in the next equation. But it is not as simple as entering all of this information into scheduling software and plotting a schedule, although some do it that way. A good detailed contract schedule will go through several versions and edits, each of which is reviewed by the PM and superintendent.

$$\text{Plan (including logic)} + \text{Durations} = \text{Schedule}$$

Many project owners and contractors prefer that a detailed schedule is attached to the contract as an exhibit. This occurs more with private than with public projects. Detailed construction schedules are often over 100 line items long with many venturing into the thousands on larger complex projects, especially lump-sum public works projects. The schedule is a construction management tool and should follow the 80-20 Pareto principle in that 20% of the activities will consist of 80% of the time, effort, and risk. It should be detailed enough that progress can be accurately measured and foremen and subcontractors can develop their three-week look-ahead schedules from it. But it should not be so detailed that management of the schedule takes on a life of its own. For instance, structural steel connections by the "connector" crew (a subset of the ironworker erection gang) should be shown on the schedule, but each beam-to-beam welded connection is not typically represented.

The detailed schedule should be posted on the general contractor's (GC's) job-site trailer meeting room wall and the GC's superintendent will provide the project owner with a schedule status update each week in a meeting. For building projects, this meeting is usually called owner-architect-contractor (OAC) meeting. Examples of a summary schedule and a three-week look-ahead schedule were included in earlier chapters and a detailed contract schedule is included on the companion website.

8.3 SCHEDULE DEVELOPMENT

The planning of the schedule and the preparation of the schedule document can be complete and independent activities, but often they overlap, as reflected throughout this book. This section breaks the actual schedule development down into additional logical steps, including process, activities, restraints, durations, time, constraints, and presentation.

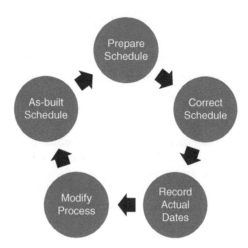

Figure 8.1   Schedule control cycle

Process

Some contractors, especially larger ones, have staff schedulers who are located in the home office. These people are experts at producing the schedule drawing, or document, with the latest software tools, but they are not necessarily the best planners. Some schedulers may have had past experience out on the jobsite, but involvement of the PM and superintendent who are going to actually build the project is paramount, just as is their involvement in preparing the detailed cost estimate. Conversely, some contractors outsource preparation of the actual schedule document. Collaboration between home office specialists and the site supervision team is the best solution.

The schedule development process is shown as only the first step in the schedule control cycle, as seen in Figure 8.1. But as reflected in this chapter's discussion, along with earlier flowcharts and an additional one in the next chapter, there are many steps necessary to prepare an accurate schedule. The best construction schedules are those that undergo many drafts and receive peer reviews – all in-house before sharing with the owner and architect and subcontractors. This process also improves buy-in from the site supervision team.

Activities

Once the project plan has been completed, it is time to develop a construction schedule. Individual tasks or activities to be accomplished were identified during the development of the WBS.

As stated, the quantity of activities in the schedule also must follow the 80-20 rule, but the nature of the project and its size will of course govern. Activities that are too short and fall outside of that rule clutter the schedule and take the focus away from the major critical path activities. Shorter-term activities, such as installing concrete formwork snap ties, if necessary, may be added to the three-week schedules. Conversely, activities that are too long also do not depict a realistic picture. For example, drawing one schedule activity line for plumbing from mobilization to demobilization is not realistic. Plumbing work can easily be divided between under-slab pipe rough-in, overhead rough-in, wall rough-in, test and inspection, insulation, setting fixtures, and trim. The description and quantity and length of activities on the contract schedule must be measurable. Industrial and heavy civil project shutdown schedules will of course again be different from this due to criticality of managing shutdowns. The scheduler has some flexibility in durations, but not too much, for example rounding 5.5 days to 6 days, but not rounding 6 days to 8.

Excessively long activities should be broken down or redefined into shorter additional activities such that they can be measurable, as reflected with plumbing and these additional examples:

- Mechanical installation could easily be split between ductwork, plumbing, and fire protection.
- Fire protection could be further split between site fire loop and hydrants, risers, overhead rough-in, trim including sprinkler heads, and testing.
- The site fire loop can be further split between layout, trench excavation, shoring if necessary, pipe bedding, lay pipe, connections and fittings, hydrants and thrust blocks, inspection, test, and backfill.
- Wood framing or rough carpentry would include breaking the system(s) down to: girders, beams, joist, posts/columns, stud packs, studs, floor sheeting, rafters, trusses, wall sheeting, roof sheeting, backing/blocking, and hardware for smaller projects.
- A large apartment project may subdivide specification division 06 by areas, such as "third floor framing," or simply "rough carpentry" for a large structural steel commercial project.

Some of these can be split even further and activities which seem too short or fall outside of the 80-20 rule can be combined into longer more summary activities. An activity worth including on the schedule should take no less than one day and activities longer than 15 or 20 days should be split into multiple activities, but both depend on the overall project duration. Some contracts limit activities to no more than 30 days. The Fire Station 83 case study project, specification section #013200.2.2/C.1 required no more than 14 day-long activities. The Army Corp of Engineers limits activities to 20-day durations. The answer is "it depends." Each project is unique. Continuing to combine activities is a method to produce a summary schedule. Figure 8.2 presents four options of longer or shorter activity durations.

After all (or at least most) of the activities have been identified, they should be assigned activity numbers in addition to descriptions. This is easily done on a spreadsheet, such as in Figure 8.3. This figure also includes predecessor and successor activities. Additional columns will be added to this series of activities later in the book.

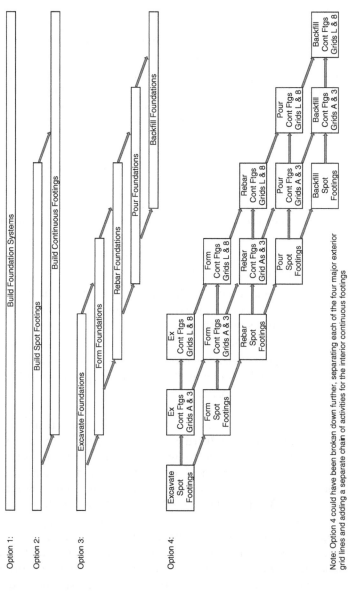

Option 1:

Option 2:

Option 3:

Option 4:

Note: Option 4 could have been broken down further, separating each of the four major exterior grid lines and adding a separate chain of activities for the interior continuous footings

Figure 8.2 Scheduling options: Long or short activities

```
┌─────────────────────────────────────────────────────────────┐
│                  Mountain Construction Company                │
│                        Fire Station 83                        │
│                          Activities                           │
│                                                               │
│   Number  Description            Duration  Predecessor(s)     │
│     70    Scaffold exterior         5           65            │
│     75    Exterior stud framing     5           70            │
│     80    Insulation                3           75            │
│     85    Dens-glass sheeting       3           75            │
│     90    Waterproof barrier        2          80, 85         │
│     95    Siding                   15           90            │
│    100    Install punch windows     5           90            │
│    105    Install exterior doors    5           90            │
│    110    Roll-up doors            15           90            │
│    115    Storefront               20           90            │
│    120    Remove weather prot.      3       95, 100, 105      │
│                                               110, 115        │
│    125    Dry-in                    0         115, 120        │
└─────────────────────────────────────────────────────────────┘
```

Figure 8.3 Activity numbers

Restraints

The next step is to determine the sequence in which the activities are to be completed. This is known as the *schedule logic*. The scheduler must consider and incorporate pre-activity restraints into the logic before formalizing and printing the schedule. Some of these considerations include contracts, permits, and material deliveries. Schedule logic incorporates activity predecessors, successors, and concurrent activities.

The schedule structure can then be developed using the relationships among the activities. Even with all of the computer sophistication available today, many contractors still develop their logic by hand on butcher paper, or by rearranging colored sticky notes on the whiteboard. The activities are moved around so that the plan represents an organized picture of how the superintendent plans to construct the project. A photograph was taken from an actual example of this from a jobsite meeting room and included earlier in Figure 5.2.

Only the first activity in a schedule does not have a predecessor or restraint and only the last activity lacks a successor. After a first plot of the schedule, a quality control (QC) check should be performed (or multiple QC checks) to make sure that there are not any dangling activities; those without predecessors and successors. The scheduler should look for errors and tie each activity to something reasonable. If there is nothing to connect an activity to, then it appears to be floating in time and therefore theoretically can occur whenever – which is very unlikely.

Durations

Direct labor is typically considered as the major resource necessary to incorporate into an early schedule. If a contractor does not have access to a reliable workforce, much of the other schedule planning effort is without value. The man-hours (MHs) a contractor has available are first determined from the detailed estimate as reflected in the planning flowchart included earlier. The transition from quantity take-off and pricing to schedule durations is partially reflected in the next equation.

> Quantity of measured work × man-hours per unit of work
> (productivity factor) = Total man-hours required
> Total man-hours / Hours per day = Total work days required
> Total work days / Crew size = Crew duration = Schedule duration

Technology today can allow the estimator to combine their database, including productivity factors, with the drawings, to produce an estimate. The scheduler can also take the next easy step and produce a schedule electronically. Restraints can be input, including resource limitations, such as a 20-carpenter crew size, and a schedule can be produced based on this input. An estimate-generated schedule is a computer function without thinking but just starting with electronic drawings, computing quantities, applying productivity factors, totaling hours, and dividing by predetermined crew sizes, without any human oversight, such as from the superintendent.

To determine schedule durations for direct work activities, the scheduler starts with the hours from the estimate, for example 130 hours to install specialties and signage. These hours, in combination with anticipated crew size, allows the scheduler to dial in on accurate durations. The superintendent's planned crew size, in this case one carpenter working fulltime and another halftime, multiplied times eight hours a day and divided into the total estimate reflects 10 work days Then the superintendent should ask, "Does this feel right?" This is referred to as the "gut check." If the duration is too slow, then the scheduler increases the crew size. After the project has been awarded to the GC, this information can easily be incorporated into a cost- and time-control work package, which will be discussed in Chapter 15.

> 130 hours estimated / 8 hours per day = 16 work days
> 16 work days estimated / 1.5 man crew = 10 work day crew duration

Most experienced estimators have in-house productivity factors for direct work gleaned from previous projects. For those that don't, *RS Means* has extensive lists of productivity factors and crew sizes. This is a valuable resource for even experienced schedulers as there are always new materials and systems that require a schedule duration. The estimate is converted to durations according to the following formulas.

Duration = Quantity / Productivity = Total hours

Duration = Total hours / Hours per day = Work days

Duration = Work days / Crew size = Crew days

Duration = 45 MBF of wall framing × 14 MH/MBF = 630 MHs

Duration = 630 MHs / 8 hours / day = 78.8 ~ 79 work days

Duration = 79 work days / 4 carpenters = 19.75 ~ 20 * crew days

*Contractors do not typically work in fractional days; use 79 and 20.

As with estimating, the scheduler should be aware of the proper use of significant digits. Work day calculations should be rounded up, for example from 4.6 to 5 days. But rounding from 4.1 to 5 might be a bit conservative, since many estimated quantities already have waste built into them. Rounding up or down according to normal rounding protocol is likely the best option. Exact schedule duration calculations also do not consider smaller assumed steps, such as set-up, tear down, cleanup, downtime, break time, or learning curve. Durations are all educated guesses or approximations.

The detailed contract schedule needs to be more than just a pretty picture hanging on the jobsite trailer conference room wall, though. Schedule durations and relationships must be realistic to be an effective tool. Overly optimistic schedules with short durations and many overlapping activities oftentimes cannot be accomplished and the superintendent will always be trying to make up lost ground. Conversely, schedules that are overly pessimistic in that durations are too long and all activities are shown end-to-end in series with little overlap are also not realistic and cannot be effectively managed in the field. If the project superintendent is ahead of schedule by two weeks, but the drywall subcontractor did not plan on mobilizing due to the pessimistic contract schedule, then the two weeks will be lost.

The best way to schedule subcontractor durations is to ask them. Subcontractor participation in the GC's schedule is a collaborative and not a top-down approach and results in buy-in from the subcontractors. This can be achieved by a subcontractor simply stating a duration, such as four weeks, or providing a schedule document. If this is not possible, then schedule durations can be figured the same way for direct labor, assuming that the prime contractor has an estimate of the subcontractor's labor hours. If this is not available, a scheduler can roughly take 50% of the subcontractor's presumed budget and assume that is for labor and divide by a wage rate and crew size to determine work days. The final option would be to use historical schedule data from similar projects and in-house resources for example, "allow two months for drywall" without considering quantity, crew size, or complexity.

Time

The duration for each activity is determined using the crew productivity factors from the cost estimate. Now the start and finish dates for each activity can be determined, and the overall scheduled project completion date is compared with the contractual completion date. If the schedule shows completion later than the contract requires, some of the activities on the critical path must be accelerated or stacked to produce a schedule that meets contractual requirements. If a duration or timeline doesn't fit with imposed constraints, the scheduler cannot simply increase crew size to reduce days. There may be tool or equipment or material resource restraints or space restraints or limits to availability of qualified craftsmen. The effect limited resources have on a schedule will be discussed in Chapter 10. The ideal situation is when a contractor can figure a way to beat the owner's schedule, as reflected in the next example.

As a general rule, partial days are not included in the detailed construction schedule. This is consistent with the 80-20 concept discussed earlier. But more detailed hourly schedules may be required for industrial projects with equipment shutdowns (called "shutdown schedules") or heavy civil projects with evening road closures. More detailed scheduling is often best handled with short-interval schedules.

A preliminary timeline is chosen by the scheduler to begin transforming the logic diagram into a schedule drawing. The exact project duration will be computed through a series of schedule calculations as discussed in the next chapter. But the scheduler can make a preliminary guess whether the project is a three-month, six-month, or one-year duration. Based on that duration a calendar must be chosen. Some of the timeline options available for the scheduler include:

Example 8.2

Most estimators can measure quantities the same, receive the same material and subcontract pricing as do their competitors, utilize the same union wage rates, and experience similar productivity in direct work – it takes two people three days to do a task. But this contractor was adept at figuring ways to build faster. If the project owner advertised the project that was out for bid would take one year, but the contractor could figure a way to schedule and build it in 11 months, they could save one month of jobsite general conditions costs and cause their firm to become the successful low bidder. Of course project success still relied on performance in the field.

- Years,
- Quarters of year,
- Months,
- Weeks,
- Calendar days including weekends,
- Work days, which exclude weekends and holidays, or
- Hours.

There may be some confusion as to what constitutes a work day. For the purpose of this discussion it is assumed contractors work Monday through Friday, eight hours a day, and 40 hours in a workweek, and do not work recognized holidays, such as Christmas and Fourth of July. But for many, four 10-hour days (four 10s) or six 8-hour days (six 8s) are also standard workweeks. But there may be exceptions to these rules of thumb. If a contractor works on Saturday, is that then counted as a work day? If they work in the rain, is it still counted as a work day? Or conversely, if it doesn't rain, what happens to the built-in rain days in the schedule?

Each work day starts at a given time, such as 7:00 a.m. (or 6:00 or 8:00), and finishes at a given time such as 4:00 p.m. (or 3:00 or 5:00). Overtime and weekend work should be reserved for adjustments and not considered in the initial plan. If an original project plan required the crew to work six 10-hour days (six 10s) per week, or seven 12-hour days (seven 12s) per week, or a double shift, there would be little room for the team to increase hours in order to make up for other schedule slips.

Some activities, such as curing concrete or gypsum wallboard taping mud, may need an adjustment for a seven-day-a-week calendar, but this may be complicated to input. Utilizing different calendars in one network, such as both five 8-hour days and six 10-hour days for different subcontractors or paths within a network is also complicated – and although the software and scheduler may be able to the accommodate that, it makes it difficult to read and use as a tool for the construction crew. The goal for the scheduler as outlined throughout this book is to allow the schedule to be a useful construction communication tool. All calculations and examples in this book are based on work days and not calendar days. Although calendar days are often the final output of the schedule document and will also be plotted. These different timeline options are reflected in Figure 8.4.

Year:							2022									
Month:			November							December						
Calendar date:	25	26	27	28	29	30	1	2	3	4	5	6	7	8	9	10
Week day:	F	S	S	M	T	W	TH	F	S	S	M	TU	W	TH	F	S
Calendar days:	175	176	177	178	179	180	181	182	183	184	185	186	187	188	189	190
Work days:	123			124	125	126	127	128			129	130	131	132	133	

Figure 8.4 Timeline options

Constraints

Constraint dates restrain logic and often are not preferred by schedulers, but may be imposed by the project owner with a required completion date. Schedule constraints are hard dates or times when something must happen. They restrict the scheduler from allowing only logic and calculations (next chapter) to solely govern. The city may restrict working days and hours when it issues the building permit. Schedule constraints include required activity or network start or completion dates, material delivery dates, equipment or road shutdowns, and so forth.

Constraint dates may also be known as milestone dates. Milestones are defined as intermediate deadlines. Examples of milestones the scheduler should incorporate into a typical commercial construction project schedule would include: Contract execution, issuance of building permits, notice to proceed, structural top-out, dry-in, energize, certificate of occupancy, certificate of substantial completion, turnover, contract close-out, and others. Specialty permits or deferred submits are often also critical and should be included in the schedule. Unique weather considerations may also be noted. The schedule logic, restraints, durations, and timeline must all be fit within these constraints.

Presentation

The final picture is then transformed into a neat construction schedule which may be used as a contract document and/or a communication tool with many members of the project team. A waterfall presentation method of a project schedule is a good communication tool. The scheduler should attempt to minimize the crossing of vertical restraint lines with horizontal activity lines. Some schedules include redundant or double-restraint lines. For example, stripping footing forms can be simply restrained by placing concrete, but technically, it is also restrained by excavation, formwork, and rebar – but these activities are understood to be restraints and it is not necessary to add vertical restraint lines between each one. Some computer-generated schedules will unfortunately have more vertical restraint lines than horizontal activity lines and are therefore not an effective communication tool. An example of this is shown in Figure 8.5. This is another example of the "horse in the barnyard" discussion from Chapter 4, Example 4.2. When drawing the physical schedule, the scheduler should avoid the overuse of vertical restraint lines. The schedule is a communication tool and if there are more vertical restraint lines than horizontal activity lines it is difficult to read. The computer simply plots the input given to it by the scheduler and does not necessarily print a readable product.

The waterfall technique focuses on completion dates first and start dates second. Another presentation is a stair-step, which is typical for fast track or overlapping activities, such as phased design and permits and phased construction starts, which are often connected with negotiated guaranteed maximum price projects more than

Figure 8.5 Activity relationships

lump sum bid projects. Two versions of a stair-step schedule diagram are included in Figure 8.6. The difference between these two is also discussed with fast-track schedules later.

Computers are valuable tools to assist in developing the schedule. The identification of the activities, the duration for each, and the sequence in which they are to be completed must be input into the computer for it to develop the schedule. The computer will plot the schedule and calculate the start and finish times for each activity. Computer-generated schedules allow the scheduler and PM and superintendent to determine quickly the effects of changing schedule logic, delays in delivery of critical materials, or adjusting resource requirements. The computer is a valuable scheduling tool, but it cannot take the place of adequate planning. The PM and superintendent should both be actively involved in developing the construction schedule, as they are the individuals who are responsible for completing the project in the desired time. This provides buy-in by the jobsite project leadership. If they agree with the schedule, they will do everything they can to make it happen.

Each of the project stakeholders needs to know the contractor's plan and each stakeholder focuses on different elements of the schedule. This includes the project owner, architect, city, subcontractors, foremen, and home office executives. After completion of schedule development, the document will be published or shared with

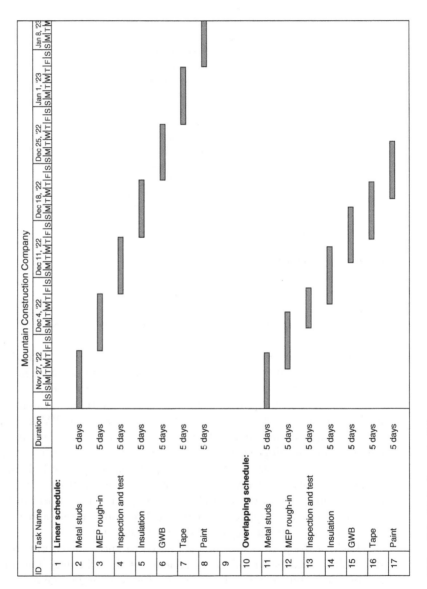

Figure 8.6 Stair step schedule

many stakeholders. Another reason the jobsite management team must publish the schedule and share it at least with the contractor's management is that in construction, it is typical that people move off the job. Superintendents and foremen and PMs may be transferred to another project or even leave the company. For some this is known as the "hit by the truck" theory in that if a field supervisor leaves, someone else needs to be able to step into position and successfully carry out the project plan.

The schedule allows the contractor and the project team to meet its long-term goals and satisfy the project owner and avoid liquidated damages. The schedule is also used to achieve short term goals as well, for example pouring the spot footings this Friday at 10:00 a.m. to allow them to cure over the weekend and strip and backfill on Monday.

After drawing the schedule, based upon the plan plus the addition of durations and dates, the scheduler (or PM or superintendent) should walk stakeholders through the schedule or a new schedule version or revision, rather than simply handing it out or mailing it or emailing it and assuming that all parties will open it and digest it. Many cannot read the schedule, but would never be willing to say so.

8.4 SUMMARY SCHEDULE

Summary schedules for proposal purposes can be produced without completion of the detailed schedule but utilizing the detail for backup makes the summary that much more accurate. Superintendents and schedulers rely on the estimated direct work man-hours. Although the estimate leads to the schedule and both lead to project controls, an early summary schedule is actually necessary to complete the jobsite general conditions estimate and to verify if the planned approach would support contractual deadlines or milestones. However, an accurate summary schedule will not be developed until after the detailed construction schedule is complete. It is therefore somewhat a "chicken-and-egg scenario"; the schedule needs to be developed before the general conditions and the estimate is complete, but the estimate needs to be complete before the detailed schedule can be developed. Summary schedules may be as short as 15–20 line items or up to 40–50 line items long. Summary schedules may also be included with responses to requests for proposals (RFPs) and sometimes they are attached to the prime construction contract agreement as an exhibit.

Bid documents often require that a schedule be submitted with the price or shortly after. Does the owner officially approve this schedule as they would a rebar shop drawing and what does approval mean? What do the specifications say? Although this is not a project management book, this author would recommend that contractors submit as many documents, samples, mockups, and so on as they can for approval. Inclusion of owner activities, such as issuing permits or delivering FF&E, should be clearly shown in the contract schedule.

A "baseline" schedule is the target schedule, which may also be the contract schedule. The baseline schedule is also known as the "as-planned" schedule, which will be compared to an "as-built" schedule in a claim discussion. It is not necessarily the summary schedule, but it is considered the first formal schedule, especially by the client.

8.5 SCHEDULE CONCEPTS

The schedule should be referenced in the *contract* as an exhibit. It should be referred to in each subcontract and purchase order by title and date. The project manager may insert language into all subcontracts placing the subcontractors on notice that they are required to achieve the project schedule and that if the GC determines that they are falling behind, the subcontractors will work overtime and/or increase crew sizes at their expense to catch up.

Commencement or day one of the schedule can be simply a date, or issuance of the notice to proceed (NTP), but neither of these stipulations are necessarily in the construction team's best interest. There are other conditions which should be in place before the clock starts for the builders, and the contract may be modified to reflect these conditions. For example, AIA contract A102, Cost-plus with a Guaranteed Maximum Price, should be modified by the GC such that Article 4.1 would read in part: The date of commencement of the work shall be . . . established as follows . . . upon receipt of:

- Verification of adequate construction financing,
- Receipt of all building permits,
- Receipt of a signed contract, and
- Receipt of the NTP.

These four requirements are often inserted into the contract. For example, ConsensusDocs contract 500 Article 6 covers this "unless modified by amendment." Most owners and general contractors will avoid using a specific date in either the commencement or completion articles of the construction contract. This was discussed extensively in Chapter 6. Completion is subsequently defined as a duration, say 365 calendar days or 250 work days after issuance of the NTP and the other aforementioned requirements.

Performing construction on a *fast-track* basis was a buzzword of the 1980's. Most construction projects are fast-track now. Basically this means that activities are overlapping and occur in parallel in lieu of in series. Very few schedules are completely linear in that the preceding activity is 100% complete before the subsequent activity starts. A simple comparison of linear versus overlapping metal stud and drywall work was shown earlier in Figure 8.6. Note both of these options also reflect the waterfall or stair-step presentation.

A project that is completed in just the right amount of time is the most cost effective. The adage *time is money* very much applies to construction. Projects which are expedited and require excessive use of overtime can potentially save on jobsite general conditions costs, but may cost more than ideal due to labor premiums and loss of productivity. Conversely, projects which are delayed and take longer than they should have will result in extended jobsite general conditions costs and equipment rental. These concepts are elaborated on more fully in Chapters 10 and 18.

8.6 SUMMARY

There is much more required to achieve project success than just scheduling, but like the cost estimate, the schedule is one of the most critical construction elements and one of the first to establish. Schedules can take on many different formats and serve different purposes. Sometimes this is predicated on the contractor's culture, potentially because of contract requirements, and likely what each scheduler is comfortable using. The contract schedule is often the detailed schedule that is displayed on the wall of the jobsite conference room and statused at the weekly OAC meeting. Three-week look-ahead schedules are developed by foremen and subcontractors and used to communicate daily activities including deliveries. A summary schedule is often required of a negotiated RFP from the project owner and typically included in executive progress reports. The original bid schedule, contract schedule, and as-built schedule are all record documents and should be kept secure for future reference.

The terms planning and scheduling are often used together. Planning and scheduling is a way for the jobsite team to think through the whole project, essentially building it mentally before any concrete is delivered. This is similar to preparing a construction estimate and many of the other preconstruction plans which were discussed in Chapter 3. Planning is the hard work that went into producing the schedule document. Planning incorporates schedule logic, deliveries, restraints, manpower, and ultimately the jobsite team's idea, or plan, on how they wish to construct the project. The schedule is prepared by inputting the results of planning into the computer and producing a document. The schedule document should be incorporated into the prime contract agreement and each subcontract and purchase order, just as drawings and specifications are incorporated into contracts. The GC's project superintendent will status, update, and revise the contract schedule during the course of construction.

On many public projects, the owner determines a start date, completion date, and work days and hours. It is up to the contractor's scheduler to fit it all in. But as can be seen from the schedule elements discussed in this chapter and others, there are many more considerations necessary to prepare an achievable schedule.

The next chapter utilizes time-related calculations to determine the overall project duration and where the *critical path* is within the network. The critical path is composed of activities that cannot be finished late or the project will finish late. Activities that are not on the critical path have *float* and can move forward or backwards, somewhat, without impacting the overall project completion.

8.7 REVIEW QUESTIONS

1. Organize these subcontractor schedule choices by preference, from first to last: (A) Use estimated subcontractor hours, (B) Call the subcontractor for input, (C) Use historical data, (D) Assume 50% labor, and (E) Just plug something in.

2. Organize these plumbing scheduling activities in logical order: (A) Ceiling pipe hangers, (B) Final approvals, (C) Under-slab rough-in, (D) Ceiling rough-in, (E) Wall rough-in, (F) Trim, and (G) Set fixtures.

3. Place these schedule development activities in logical order: (A) Print the schedule, (B) Prepare a WBS, (C) Detailed estimate, (D) Prepare an as-built schedule, (E) Plan the schedule, and (F) Status the schedule.

4. What is the difference between scheduling and planning?

5. Which members of the construction team input to the project plan?

6. Are schedules contract documents?

7. When should day one of the contract schedule occur?

8.8 EXERCISES

1. Place these potential sources for productivity factors in relative preferred order: Superintendents, foremen, subcontractors, databases, in-house specialists, company history, and the scheduler. There is no exact answer.

2. How many activities should be included on the project schedule?

3. Place these overhead construction activities in logical order: (A) Ceiling grid, (B) Pipe insulation, (C) Cover inspections, (D) Mechanical, electrical, and plumbing rough-in, (E) Label valves and pipe runs, (F) Pressure test plumbing and fire protection, (G) Install ceiling tiles, (H) Pipe hangers, (I) Heating, ventilation, and air conditioning balancing.

4. Provide three examples of construction activities that fall both within and outside of the 80-20 rule for the fire station case study project.

5. Utilizing the detailed schedule from the companion website, develop a schedule unique to the electrical contractor. Include work of other trades as it affects the electrician's work.

6. Develop a short interval schedule for the construction of the exterior wall and siding work beginning with stud installation for the case study project. Identify the activities and logic, and take a shot at reasonable durations for the activities. Include at least twice the quantity of activities included on the detailed schedule.

7. From among the following list of activities, with predecessors shown, draw a simple logic diagram.

Number	Description	Predecessor(s)
10	Notice to proceed	
12	Demolish existing structure	10
14	Survey utilities	10
16	Install erosion control	10
18	Clear and grub site	12, 16
20	Mass excavation	18
22	Import and place select fill	20
24	Trench site utilities	14, 22
26	Pipe bedding	24
28	Lay pipe	26
30	Pipe fittings and restraints	26
32	Hydrants, CBs, MHs	24
34	Test	28, 30, 32
36	Inspection	34
38	Backfill	36

8. Add a generic two-day duration to each activity (except 10, which is zero) in the logic diagram prepared in Exercise #7 and draw a network schedule. What is the list of activities the critical path flows through?

9. Refer to Exercise #8 above. Prepare a new schedule where activity 12 is changed to one day, 20 and 22 are both changed to four days, 30 and 32 are changed to five, and activities 34 and 36 are changed from two days to one day. What is the new list of activities the critical path (s) flows through?

10. Add the following durations to the logic diagram created from Exercise 13 in Chapter 4. What is the overall length of this network? Activity 10: 1 day; 20: 2; 30: 6; 40: 8; 50: 2; 60: 5; 70: 4; 80: 12; 90: 2; and 100: 2.

Schedule Calculations

9.1 INTRODUCTION

Today computer scheduling software programs perform all of the scheduling calculations or schedule mechanics. Most often the calculations and their results are hidden from the schedule output. Most users of the project schedule are unaware of the results of the calculations, other than that activities on the critical path are shown and differentiated from noncritical activities. But because this is a book dedicated solely to scheduling, it is relevant that the process of performing the calculations be discussed along with a few examples and exercises. This way the reader, and one day "scheduler," will have a better understanding of why one activity has been shown as critical and another not.

All built environment participants should be aware of a few common technical scheduling terms and concepts. Many of these terms are described in detail in this chapter and are also included in the glossary. Schedule calculations involve four basic steps, which comprise the basis of this chapter. These steps include:

1. Forward pass, including early start and early finish dates;
2. Backward pass, including late start and late finish dates;
3. Float, including total float and free float; and
4. Critical path.

These calculation steps, and others, occur in order, as reflected in Figure 9.1. Also included in this chapter is a brief precalculation refresher section overviewing logic, relations, durations, and timelines. This chapter is a continuation of the logic prepared in Chapter 4 along with the durations and timeline established in Chapter 8.

Precalculation Refresher

Chapters 4 and 8 on schedule planning and development respectively have laid the groundwork for this chapter on calculations, including determination of the critical path. The main points from those chapters and others have formed the important foundation for creation of the schedule. This chapter, and subsequent chapters, including Schedule Control in Chapter 13, will build on that foundation. Before the schedule document is prepared the contractor should go through a vigorous planning effort. Planning incorporates all of the project activities, often with help from

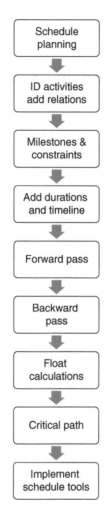

Figure 9.1 Schedule calculation process flowchart

the work breakdown structure. Those activities are then arranged in a logical order. Relationships between activities must be identified. Relationships are also known as dependencies or restraints. Essentially relationships identify what needs to happen before one activity can happen or what happens immediately after an activity is completed.

Constraints are differentiated from restraints but are often discussed in the same vein. Constraints are dates often imposed by others on the contractor that can influence the schedule. Constraints include everything from a notice to proceed (NTP) date to a hard required completion date. Permit issues and material delivery dates must also be incorporated into the schedule. Each activity has a defined start date, duration, and end date. The focus of this chapter is calculating those start and completion dates.

The durations are often influenced by the cost estimate, in particular the direct work man-hours, combined with crew sizes, for each activity.

The timeline for a schedule is the total duration the project will take to build. The scheduler has a variety of timeline options to incorporate into the schedule document from years and months to weeks, days and hours. The proper choice of timeline depends upon the length and complexity of a project.

9.2 FORWARD PASS

The first set of calculations performed when analyzing a schedule network to determine the critical path is the forward pass calculation. This process determines when each activity in a network starts (its first event time) and when it ends (its second event time). The start and end dates calculated by the forward pass are known as each activity's early start (ES) and early finish (EF) times. These are the earliest dates any activity can occur as determined by a combination of the network's logic, relationships, and milestone and constraint dates. The end result of adding each activity's durations together, adjusting for nuances with logic and relationships, is the overall duration of the project and its scheduled completion date.

The forward pass can be performed with either an arrow diagramming method (ADM) or a precedence diagramming method (PDM) network format. Graphically, for the scheduler it is easiest to prepare a schedule that logically is laid out from left to right and from top to bottom. The activity on the far left and at the top of the schedule is typically the first in a network. The activity at the far right and bottom of the network is typically the last activity. This was the waterfall effect presented prior.

The forward pass calculations first require that the logic and associated activity relations and restraints are complete and correct. The forward pass also requires a duration (days or hours or weeks) for each activity and incorporates milestone and constraint dates. The first date needed is the network start date, which for many construction contracts is the official NTP issued by the project owner. Combining these important dates, a calendar based on work days, and proper logic and activity durations, the scheduler proceeds with the forward pass, the first step in determining the critical path. The basics of the forward pass calculations are shown in the next equation and can best be explained with another series of steps, as shown in Figure 9.2 and Table 9.1.

$$\text{ES (Activity 1)} + \text{Duration (1)} = \text{EF (1)}$$

$$\text{EF}(1) = \text{ES}(2)$$

$$\text{ES}(2) + \text{Duration}(2) = \text{EF}(2)$$

$$\text{EF}(2) = \text{ES}(3)$$

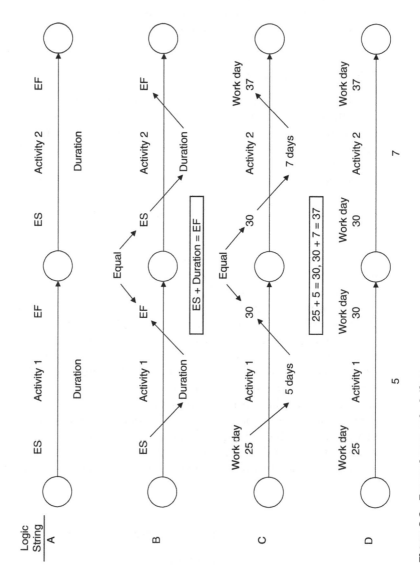

Figure 9.2 Forward pass calculations

Table 9.1 Forward pass calculations

Activity				Start	Finish
#	Description	Duration	Predecessor	Day #	Day #
10	Excavate	2		1	2
15	Formwork	5	10	3	7
20	Install rebar	2	15	8	9
25	Anchor bolts	1	20	10	10
30	Place concrete	1	25	11	11
35	Strip forms	1	30	12	12
40	Backfill	2	35	13	14
	Total duration:	14 work days			

Notes: 1. It is assumed that all activities start at 8:00 on the start day and finish at 5:00 pm on the finsh day.

2. The forward pass calcuations begin with the first activity in the network.

For this example, it is assumed that activity 10 begins on day 1.

3. The start day # is the morning after the predecessor's finish day.

4. The finish day # is calculated by adding the duration of the activity to the start day #, but allowing for both the start and finish days to be included in the duration.

5. Day # are work day numbers, not calendar days.

6. The start day # is the early start date from the forward pass.

7. The finish day # is the early finsh date from the forward pass.

There are two starting day conventions when performing schedule calculations. The first option starts the project's first activity on day zero for the first day of the network. The day zero method uses the start time as the end of the previous day (5:00 p.m.) and the completion time as the end of the first day, day one. The second is known as the day one method, which uses the start time of this activity as the beginning of this day (8:00 a.m.) and the completion time as the start of the next day, day two. Using the day one method is known as "the beginning of the day convention" and is the one used for examples in this book. The start and finish dates for each activity and the entire network can then be calculated by the sample addition formula noted here.

> Day one + Activity duration(s) + Predecessors and successors = Cumulative work days, and later calendar dates

Constraint dates can either be placed early into the logic diagram, or early schedule draft, or after calculations have been completed. If the scheduler waits to impose constraint dates until after the schedule calculations are complete, he or she is starting the network with a best case scenario. If constraint dates govern after input into the network, they are then the critical path and activities or paths will need to be shortened or rearranged to accommodate these constraint or milestone dates.

For the calculation examples in this chapter an arrow diagramming method or activity on the arrow (AOA) network will be used. Activity on the arrow network was the first network method, popular in the 1970s, but replaced by many schedulers in favor of PDM, or activity on the node (AON) networks. A PDM network example will also be presented later in the chapter.

The forward pass determines the project duration. Production of the schedule could be complete after the forward pass is done and printed and float calculations ignored. A schedule prepared without displaying float is a strategy discussed later. Many contractors and even some schedulers prefer not to show any float.

Very few, if any, schedules happen with each activity occurring in sequence as reflected in Table 9.1. For most, if not all construction networks, multiple activities occur relatively in parallel. The objective of the calculations discussed in this chapter is to determine which of these are critical activities, and which path(s) of activities represents the critical path(s). The following three steps need to be added to the forward pass steps in Table 9.1 when parallel activities occur:

1. When two or more activities finish at the beginning event node (i node) of a new activity, the latest of the early finish dates of the preceding activities is used as the early start of the new successor activity.

2. When two or more activities start from the same beginning event node (i node), each utilizes the early finish of the precedent activity (or latest EF if there are multiple precedent activities) as their early start date.

3. The duration of each of the new parallel activities is added independently to their respective ES dates to determine each of the subsequent activities' EF dates. These steps are then followed for the remaining activities in the network.

9.3　BACKWARD PASS

It is important for the scheduler and the builder to know the early start and finish dates as established in the forward pass. It is also important for the scheduler to know the latest these activities can start and/or finish without delaying the completion of the project. These are known as the late start (LS) and late finish (LF) dates, respectively. These dates are determined by the backward pass calculation.

The backward pass establishes the criticality of each activity. The backward pass is the second set of calculations to determine the critical path of a project network. The backward pass is essentially a reverse function of the forward pass. The scheduler starts with the last activity in the network, which is logically the one on the far right and at the bottom of a waterfall schedule presentation. To establish the late event times of a network the late finish time of the last activity (its j node) is set equal to the early finish time of that activity. Calculations are then performed in reverse of the forward pass, starting with each activity's completion date and subtracting its duration, to determine the activity's late start and finish dates as reflected in the next calculation and the detailed steps shown in Figure 9.3 and Table 9.2. The backward pass determines the late start and late finish dates of each activity. These are the latest event times that each activity can occur without affecting the project's critical path and the overall completion date.

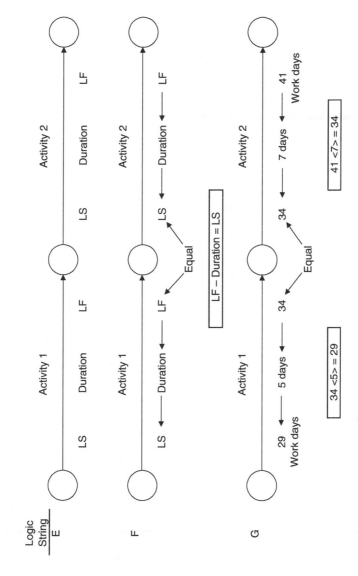

Figure 9.3 Backward pass calculations

$$LF(Activity\ 3) - Duration(3) = LS(3)$$

$$LS(3) = LF(2)$$

$$LF(2) - Duration(2) = LS(2)$$

$$LS(2) = LF(1)$$

Table 9.2   Backward pass calculations

Activity #	Description	Duration	Predecessor	Finish Day #	Start Day #
40	Backfill	2	35	14	13
35	Strip forms	1	30	12	12
30	Place concrete	1	25	11	11
25	Anchor bolts	1	20	10	10
20	Install rebar	2	15	9	8
15	Formwork	5	10	7	3
10	Excavation	2		2	1
	Total duration:	14	work days		

Notes: 1. It is assumed that all activities start at 8:00 on the start day and finish at 5:00 pm on the finsh day.
2. The backward pass calculations begin with the last activity in the path.
3. The finish day # of the last activity in the path, which is the starting activity in the backward path calculation is the same evening of its early finish day dermined in the forward pass.
4. The start day # is calculated by subtracting the duration of the activity from the finish day #, but allowing for both the start and finish days tobe included in the duration.
5. Day # are work day numbers, not calendar days.
6. The start day # is the late start date from the backward pass.
7. The finish day # is the late finsh date from the backward pass.

The late start and late finish dates for the example in Table 9.2 assumed that each activity was sequential, similar to Table 9.1, but as stated, this rarely happens in construction. Most projects have multiple paths of activities with multiple activities occurring in parallel, or partially in parallel, along each path. The following three additional steps are necessary when activities occur in parallel.

1. If more than one activity follows or succeeds another activity, the successor activity with the latest LS date is used as the LF of the preceding activity.
2. If multiple activities occur in parallel, the LF of each of the preceding activities will be the same as the LS of the successor activity (or latest LS if multiple successor activities).
3. The LS of each of the parallel activities is determined independently by subtracting each activity's duration from their LF dates.

These steps are then followed for each preceding activity, moving from right to left in the network, and finishing at the original starting activity or node or event time, often the issuance of the NTP. The backward pass must finish at the left side of the schedule on day zero; if not, there is an error the scheduler must resolve for. Both the ES and EF of the first activity will be the same, in this case zero, just as the EF and LF of the last activity are the same.

9.4 FLOAT

After the completion of the forward and backward pass calculations, the third calculation necessary to determine the critical path is for float. Float calculations begin with the result from the backward pass, which calculates late start and late finish dates. Float is the difference between late start and early start dates and/or the difference between the late finish and early finish dates. The result of either of these simple calculations for each activity should be exactly the same. If the calculation of the difference is not the same, then there is an error and the scheduler must resolve for this.

Since the critical path does not include all the activities on a detailed schedule, the term *float* refers to additional time available for certain activities, or a string of activities, not on the critical path before these activities move into a new critical path. For activities outside the critical path, their duration can slip, either in start or end dates, but the project can still be completed per the contract schedule. However, it is also true that delaying these activities beyond their allowed float would cause them to become critical.

Float is also known as contingency time or slack, although the term *slack* has a negative connotation, for example, there cannot be any "slackers" on a construction site. Float represents flexibility in the schedule. The float of one path is determined by the longest float of any one activity along that path.

Total Float

To determine the critical path, a scheduler performs first a forward pass and then a backward pass through all of the paths in a project network. As discussed, the forward pass calculates the early start and early finish dates for each activity. The backward pass calculates the late start and late finish dates for each activity along each path. The difference between the two sets of dates (early versus late) is known as total float (TF). The total float values are not difficult to compute as reflected in the next equation.

> LS – ES = Total float, which should be the same as LF – EF, for any one activity

Any activity that does not include float, or has zero float (LS – ES or LF – EF = 0), is considered "critical." When the total float of an activity or path of activities

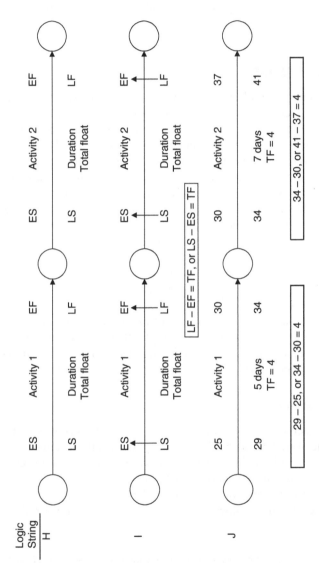

Figure 9.4 Float calculations

or total network equals zero, there is not any flexibility in the schedule. To avoid delaying the project, any activity with zero float must be started as scheduled, take no longer than the duration it is scheduled for, and finish no later than its scheduled completion date. This is true for every activity and every path in the network that has zero float. The TF calculations for the previous two figures are depicted in Figure 9.4.

Free Float

Free float (FF) is defined as the maximum delay available for one activity along a path without delaying the early start of a successor. Free float does not delay project completion. Total float is more commonly used in construction scheduling than is free float. If an activity has both TF and FF, the TF will be greater than the FF. Free float is the amount of time one specific activity can be delayed within a path of activities but the delay of that activity will not affect the others; unless delayed beyond its allowable free float time. Free float may also be known as activity float. For all critical path activities both the TF and FF equal zero.

If the calculated project completion date is greater or longer than the contract date, the network has *negative float*. At that time the scheduler needs to refigure the network or the contractor may choose to pass on the job and not proceed with a bid. If the calculated finish date is less than or earlier than the contracted date, then the contractor has a time contingency or float. The contractor may choose to schedule the project based upon the earlier date, which creates an opportunity for the estimator to reduce jobsite general conditions cost and potentially realize a lower bid.

The term *negative float* indicates that there are fixed constraints in a network and at present there is not a path or plan to get the project finished on schedule. Contractors do not plan a project with negative float and will work to find solutions to resolve problems in the network. It would be unusual to find a printed construction schedule with negative float shown. Negative float raises a flag to the scheduler indicating either there is a problem in the logic or a revised path or plan is necessary.

Dummy activities are included in an arrow diagramming networks to show relationships and distinguish between two activities which may have the same ij node numbers. Dummy activities do not have float or durations, but they do have early and late start and finish event times. Precedence diagramming networks do not have a need for dummy activities. Dummy activities were shown previously in Figure 7.7.

Although the arrow diagramming method (ADM) was used for examples in this chapter to show calculations, some schedulers prefer to use the precedence diagramming method (PDM). Figure 9.5 is the PDM version of the end result of the string of ADM calculations; line J in Figure 9.4. One advantage of the PDM is the ability to include additional information about each activity on the schedule. Some additional items that can be added to a PDM node include the percent complete, original duration, remaining duration, resources, responsibilities/subcontractor, and others. But too much information included on a node may make that schedule unworkable.

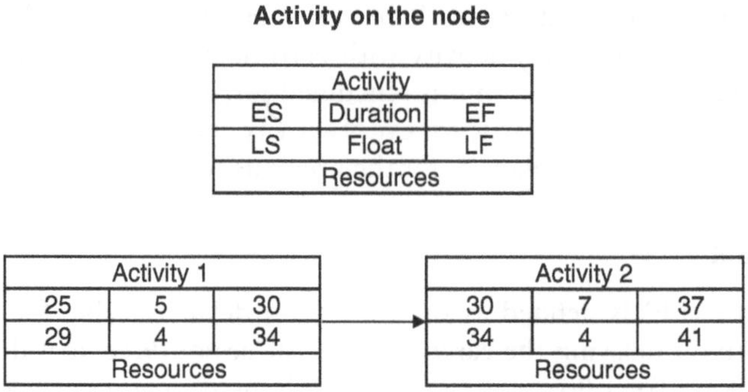

Activity on the node

	Activity	
ES	Duration	EF
LS	Float	LF
	Resources	

	Activity 1	
25	5	30
29	4	34
	Resources	

	Activity 2	
30	7	37
34	4	41
	Resources	

Note: The interior six cells of each node can be set up as formulas utilizing MS Excel.

Figure 9.5   Precedence diagramming method calculations

Strategies

If the contract date or constraint date is later than the scheduler's calculated finish date then the entire network has float. The scheduler can incorporate a couple of different strategies with project float. One is to move all activities to their late dates, which then uses the entire float up and does not provide the builder with any flexibility to incorporate changed conditions. This strategy also relaxes schedule "criticality" in the minds of subcontractors; they then may delay their mobilization. Another approach is to calculate float from a late constraint date, which technically means all activities in the schedule are then not critical. This also makes for a difficult planning and control effort for the general contractor's (GC's) project superintendent. Another option is to use the ES and EF for the entire schedule, which is the most optimistic and expedited schedule, and reduce the jobsite general conditions estimate.

If the calculated finish date is later than the contract date then the network has negative float and must be resolved either before the general contractor submits its bid or enters into a contract, especially one with liquidated damages. If the negative float is discovered during the course of construction, alternative means and methods must be employed to draft a recovery schedule. Negative float may also be known as "lead."

Who owns the float? The contract usually specifies this critical issue. Usually, it is owned by the first party who uses it. The project owner may claim the float is theirs and use it to introduce new scope. The architect or engineer may use some of the float when responding to requests for information and submittals. Subcontractors and suppliers may use some of the float and either start their work later or deliver materials later. Many built environment participants have the first come, first served mentality when they see float in a schedule.

There have been numerous debates over float ownership, including many resolved by the legal system; therefore, like the discussion of top-down scheduling, float ownership may be a contentious issue as well. The scheduler would be well-advised to study the project's special conditions regarding float. A formal dispute over float ownership will be resolved by the contract, state law, or case law. If a scheduler wishes to remove or hide the float, he or she has a few options, including:

- Utilize only ES and EF dates: This gives the scheduler more flexibility in dealing with "who owns the float" questions.
- Utilize only LS and LF dates – this is risky as all activities are then critical.
- Add additional constraint dates.
- Show relationships or additional restraints.
- Add resource requirements, which are also restraints.
- Add new activities: The scheduler can insert additional smaller ones, which are often understood to be included in longer activity durations, such as sort or unload materials, stage, prep, layout, start, continue, test, inspection, equipment start-up, cleanup, finish, and others. Examples of this are included in schedule fragnets throughout this book.
- Inflate the duration of activities and hide the float.
- Look real hard: Is there nothing actually occurring during this time period?

There are many more technical analyses of float and schedule calculations beyond the scope of this book and more geared for production industries or professional schedulers or construction claims consultants. The interested reader may look to some of the supplemental reference materials located at the back of this book. Some of these concepts and terms that are not common in construction but are included in a larger scheduling vocabulary include:

- Independent float: Owned exclusively by one activity, not available for use of any other activity or path of activities; also known as "safe float."
- Interfering float: Held by two or more activities and potentially a path of activities, but not multiple paths and not the project; also known as "shared float."
- Start-restraining float.
- Contractor-created float.
- Finish-restrained float.
- Double-restrained float.
- Unrestricted float and others.

The existence of float in a construction schedule is a good thing for the general contractor's project superintendent. Float allows the builder flexibility that does not exist with activities on the critical path. This flexibility can be used to start an activity at a different time, take longer than was determined as the original duration, and/or finish at a different time. This flexibility in the schedule can be used by the superintendent to:

- Level resources, such as manpower or equipment use.
- Incorporate minor changed conditions.
- Incorporate restraints, such as material deliveries, that were not shown on the original schedule.
- Work with subcontractors to ease their schedule constraints, and others.

Some feel that the home-office staff scheduler should determine which strategy to use with respect to float; whether to show it or not or which activities can consume it. It is this author's opinion that the project manager and the contractor's office will strategize on float management with respect to the project owner and design team. The GC's project superintendent should be the one who determines how float is handled as it relates to subcontract management.

9.5 CRITICAL PATH

The critical path consists of a sequence of activities or paths within a network, between the first activity and the last, for which total float equals zero. The critical path is also known as the critical path method or critical path network. The critical path has already been determined by the three previous steps. This is not a step in the calculation process per se, but determination of the critical path is the answer that the three previous steps produced.

A construction path is defined as a continuous set of related activities. A path of activities is also known as a chain. The longest path is the critical path. Any activity that is not critical has float. A network diagram combines all activities on one schedule. It is a collection of activities all related to one project. There are usually many paths in a network. If any activity actually starts or finishes on its late dates it now becomes critical, which may have an effect if a new path becomes critical and requires a recalculation of the entire network. The construction schedule may then have a new critical path, or several new critical paths.

After the math is done, a superintendent should become involved to review for gut checks. Are these relationships and restraints correct? Are activities missing? Do the durations appear reasonable? Is this string of activities truly the critical path? Most of the construction activities in the following list are critical in a typical commercial construction project. If these or other similar activities do not show as critical, that is, they have float, the scheduler should check again.

- Mass excavation and fill,
- Substructure concrete, including foundations,
- Superstructure concrete or masonry,
- Structural steel and/or rough carpentry framing,
- Roofing,

- Building enclosure, including curtain-wall,
- Switchgear delivery and energizing,
- Elevator delivery, installation, and buyoff,
- Mechanical and electrical rough-ins and cover inspections,
- Drywall and others.

The scheduler prints drafts and edits the schedule before distributing to other stakeholders. He or she should look for common scheduling errors, which are not necessarily a three- versus four-day duration judgment but significant errors, such as the following:

- Missed activities, such as deliveries and permits and submittal approval;
- Missed restraints;
- Improper restraints;
- Incorrect logic: The scheduler, if not also the builder, should review the draft logic (and durations) with a builder such as the superintendent;
- No buy-in from owner, architect, superintendent, subcontractors, staff scheduler, and/or project manager;
- Creating the schedule in a vacuum, top-down versus collaborative scheduling;
- Too much detail: Remember the 80-20 rule;
- The scheduler should know his or her audience, that is, know the user. The superintendent does not need or require the same information the GC's chief executive officer does or the architect does. Each member of the team, including subcontractors, has different experiences with reading and interpreting schedules and is looking for different information from the schedule.

9.6 SUMMARY

All of the schedule calculations are performed by scheduling software today, but it is relevant for the scheduler and builder to understand how the critical path was calculated.

Before the schedule mechanics are performed, it is imperative that all of the activities are identified and arranged in a logical order, including showing all relationships. After completion of a variety of planning input steps, the scheduler incorporates activity durations and plots the network according to a timeline.

The forward pass calculations determine the early start and early finish dates for each activity and the shortest duration that the project can be built. This is essentially adding the duration of each activity to the ES date which determines the EF date. The backward pass mechanics determines the late start and late finish dates and is performed in just the opposite fashion of the forward pass. Each activity's duration is subtracted from its LF date which determines its respective LS date. The total float is the difference between the LS and ES dates, which should be the same as the difference between the LF and EF dates. If the LS – ES and/or LF – EF dates are greater than zero, the activity has float. If they are equal to zero, they are deemed to be critical activities. The critical path is the string of activities that individually and collectively have zero total float.

After completing all of the calculations, the total project duration, and finish date, must be compared to the contract date. If the project is scheduled to complete after the contract date, which is also a constraint date, the network has negative float and the scheduler must find an alternative approach or the contractor may choose to pass on the project, especially if it has liquidated damages. There are a variety of strategies a scheduler and builder can take, whether they show or hide the float in a schedule network.

After drawing the schedule and performing all of these calculations, the scheduler should walk the jobsite team (project owner and superintendent) and other project stakeholders through the new or revised schedule, rather than simply handing it out or emailing or posting it to a website or drop box. The scheduler cannot assume that all of the parties will download and print and digest the schedule, especially if it has hundreds or even thousands of activities. Performing all of these calculations may produce accurate numbers, but the schedule must be realistic and must receive buy-in from those who are going to use it as an effective communication tool.

9.7   REVIEW QUESTIONS

1. Why should the GC scheduler bother with float strategies and not simply publish the schedule as is with float shown throughout?

2. Are activities 1 and 2 in Figures 9.2, 9.3, and 9.4 on the critical path?

3. Which schedule calculation process determines (A) total project duration, and (B) float?

4. If one activity or path shown on the critical path slips by one day, what happens to the project completion date?

5. If the day zero convention is used in lieu of the day one convention, will the entire network be finished one day quicker?

9.8   EXERCISES

1. Why is it imperative the scheduler get the logic and relationships correct before applying durations and a timeline?

2. If a potential project with a $10,000 per day liquidated damages clause appears to have 20 days of negative float, and the scheduler cannot resolve it, what might the contractor do?

3. Insert activity 3 into Figures 9.2, 9.3, and 9.4, which is restrained by activity 1 and has a finish-to-finish relationship with activity 2. The duration of the three is 12 days. (A) What is the critical path of activities? (B) Which activities have float? (C) What is the new EF date of this string of activities?

4. In lieu of Exercise 3 above, insert activity four into the calculations in Figures 9.2, 9.3, and 9.4. Activity 4 is restrained by activity 2 and has a duration of one day. The commencement of activity four is constrained by work day 35. Without adjusting schedule logic or activity durations, what is the total completion of this string of activities and what are the revised float calculations?

5. Revise Tables 9.1 and 9.2 by incorporating the following changes: Formwork lasts three days, concrete placement occurs over two days, and backfill lasts three days. What is the new duration of this string of activities?

6. Ignore Exercise 5 and calculate a new duration for Tables 9.1 and 9.2 incorporating these changes: Formwork begins one day after excavation starts, rebar begins three days after formwork starts, anchor bolts are inserted the same day as the concrete is placed, and backfill starts the same day as form removal. What is the new duration of this string of activities?

7. In addition to the list of potential critical activities provided earlier in this chapter, what are examples of critical activities in a typical residential, commercial, and/or civil project?

8. Did the project schedule posted to the jobsite trailer conference room wall on your project show any of these dates or calculations: ES, EF, LS, LF, FF, or TF? If not, did these calculations not exist? If they did exist, but were hidden, why was the schedule displayed in that manner? Was it an ADM or PDM network diagram or possibly a simple bar chart?

9. Calculate the ES, EF, LS, LF, and TF for the schedule developed from Exercise 10, Chapter 8.

Chapter **10**

Resource Balancing

10.1 INTRODUCTION

Contractors need a variety of resources to build construction projects. Although it is understood they need labor and material, there are often other resources necessary, all of which have direct connection to schedule development. This chapter discusses several of those construction resources and ways schedulers and field managers will work to balance available resources within the parameters of the project schedule. Resource planning and management was introduced in Chapter 4 with the initial schedule plan.

Resources include direct labor, construction materials, temporary construction equipment, subcontractors, cash or financing, field management personnel, and corporate organization support. Cash is a resource in construction, as it is for all industries, but especially because contractors typically are not paid until well after work has been put in place and they have spent their own money out of pocket. Cash flow scheduling is an important resource and will be discussed in the next chapter. Even time is a resource. The basic estimate equation presented in Chapter 2 is repeated here. Each of these elements are also resources which require management by the contractor.

Direct labor

+ Direct material

+ Construction equipment

+ Subcontractor costs

+ Jobsite general conditions

+ Markups including fee

= Construction cost estimate

Each of these estimate categories requires resource management.

- Direct labor for the general contractor (GC) includes carpenters, laborers, iron-workers, and other trades.
- Direct materials comprising concrete, steel, rough carpentry, doors and finishes.
- Construction equipment includes a forklift, crane, pickup truck, compressors, welders, and others.
- Subcontractors involve mechanical, electrical, roofing, and many other specialty companies.
- Indirect labor includes the project manager (PM) and superintendent.
- Markups including liability insurance and bonding capacities.

The discussion in this chapter is focused on limited resources, not unlimited resources. For example, what if a small town had access to only one tower crane and it was tied up on a two-year project, but a new seven-story apartment was planned that also needed a tower crane? An additional tower crane would need to be found or the apartment project delayed or other resources put into play. Each of the resource categories included in the estimate equation above are limited resources.

Just as an as-built estimate is important for estimating future work, an as-built schedule is helpful in scheduling future projects. Similar to estimating, the greatest risk in scheduling is the determination of the direct craft workforce productivity. The jobsite team and home-office scheduler should develop a set of productivity factors based on actual prior experience. An as-built schedule can help with this. These productivity factors will help the PM and superintendent establish realistic activity durations when scheduling future projects. Labor productivity, including overtime affects, is also discussed in this chapter.

With respect to resource management, and other scheduling applications, this book is not theoretical and not based on algorithms and other theoretical formulas, but rather construction industry tools. Scheduling software and complicated formulas can produce a variety of histograms, schedules, and tables to plan and monitor resource tools. As stated elsewhere, the scheduler needs to know their audience and produce documents that assist with building the construction project. Reports generated by the construction scheduler are also introduced in this chapter.

10.2 RESOURCE ALLOCATION

Schedules should be developed based on logic, relationships, and durations first, and then the scheduler should analyze how the schedule fits with available resources. Some advocate utilizing resources to first drive the schedule and manipulate the logic and timeframes to support a balanced use of resources. This feels like a backwards approach as it ignores workflow and craftsmen's productivity and well-being. A schedule predicated on resource leveling over logic is plagued with activity interruptions due to multiple start and finish dates. Driving the schedule first based on resources can be accommodated by scheduling software. This is an academic research approach which relies on the use of complicated formulas and algorithms, but this book's focus all along has been developing a schedule with the builder in mind.

There are many variables involved with construction productivity including quality, safety, experience, learning curve, equipment, tools, supervision, temporary power, and even crew parking. All of these are relevant and should be incorporated into the schedule. Allocation of resources can be added to a network as restraints or new activities. Some resources may be added as relationships, such as delivery of the tower crane by a certain date. Additional nonresource information can also be added, such as contact information (people, phone, emails) of subcontractors and suppliers. This is good information for a superintendent to have, but the schedule may now become lost among the resources, similar to the "horse in the barnyard" (Example 4.2).

Resources should not all be averaged for the entire project, or distributed in a straight line; rather, they should be assigned proportionately to each activity. Activity-based costing (ABC) is an approach to track, identify and assign indirect costs to direct work activities such that the total cost impact of that task may be known. This tracking feeds into lean construction, which then focuses on reducing the costs of inefficient activities. The ABC approach applied to allocating resources to a construction schedule may now be known as activity-based resourcing, or ABR. Some resources may not be allocated 100% to activities. For example, costs of construction equipment, such as a forklift or pickup truck or tower crane, and indirect manpower, such as the superintendent or project engineer, must be spread across the entire project.

Scheduling software allows resources to be loaded to the schedule. Necessary resources are applied to individual activities or can be spread, such as hours, crews, manpower, and equipment. But if the scheduler places too much emphasis on resource loading the schedule, some of its effectiveness as a communication tool may be lost. If the scheduler adds cost to the schedule it also loses its first focus as a tool to help build the project. The schedule is not a pay request but it can be used to support the pay request process as discussed later.

10.3 BALANCING, NOT LEVELING

This book promotes a reasonable resource balancing plan. This approach differs from forcing the schedule into a perfectly level use of resources, which for a variety of reasons, may not be practical or cost efficient. Forced balancing of resources allows resource utilization to be more uniform and more economical, but it may also use up some of the float and/or impact the critical path. The scheduler cannot cause potential resource utilization to be completely flat or level without sacrificing some efficiency.

The goal of managing resources is to smooth the curve, that is, "balance" but not necessarily "level" resource needs. The scheduler cannot achieve a perfectly flat curve without seriously sacrificing the schedule. Most resources cannot be perfectly "leveled" flat but many can be relatively "balanced." There will always be some peaks and valleys. Larger projects have more flexibility with this than do smaller projects. Negotiated projects also have more flexibility in resource balancing than do competitive bid lump-sum projects. Moving noncritical activities forward or backward to better utilize resources, such as construction equipment and manpower, is often achievable and desirable.

The scheduling software can move activities around with the sole goal of leveling resources, but would a builder want this? The software is not the scheduler, and schedule logic as determined by field management must take priority. A scheduler cannot simply move construction activities around such that a perfectly level resources curve is established. Critical activities cannot be delayed. Activities that have float may be able to move, but most/all activities have restraints and moving any activity left or right, earlier or later, just to balance resources must consider both upstream and downstream schedule affects.

It has been stated by others that the general contractor does not need to concern itself with balancing manpower if it uses subcontractors. This is not true. The total manpower needed on a jobsite will be relatively the same, whether the GC is performing concrete formwork or steel erection with its own crews or employing subcontractors. Manpower is a limited resource and craftspeople often work within a limited space and for limited hours per day, as discussed later in this chapter. Subcontractor manpower requirements are especially relevant if the GC views its subcontractors and their employees as crucial team members, and not as commodities. The collaborative approach to scheduling subcontractor work is discussed throughout this book and is relevant to subcontractor resourcing as well.

Construction schedules are typically not created or restrained by labor leveling concepts, unless there is a lot of float, which is unusual. Creating such a schedule would be known as a "resource-driven schedule." There are additional terms outside the scope of this book with respect to resources including forward resource leveling and backward resource leveling. Those concepts will be left to a more advanced scheduling study.

10.4 LABOR PRODUCTIVITY

Planning for, and managing labor productivity is more an art than a science. There are many labor productivity factors in publications like *RS Means* and the *Guide*. Experienced estimators and schedulers will have their own in-house database with productivity factors saved from as-built estimates and schedules. But how can a contractor be sure the crew will achieve these rates? A previous construction executive once told this author: "I don't mind if the competition sees our labor productivity rates. They cannot achieve the level of productivity as we can with our guys and gals proudly wearing our company logo on their construction gear."

Many in construction utilize the term *parade-of-trades*, which means one trade, or specialty contractor, such as the floor covering craftsmen, follows the work of the painters, which follows the work of the drywallers, which follows the work of the insulators, and so on. A mid- to high-rise apartment building would experience this parade-of-trades and a natural balance of labor would occur.

When developing crew sizes and durations, there are many factors that need to be considered and input by the project team, especially the superintendent. It is not simply an automated calculation determined from the estimate. Some of the schedule considerations that must be factored include: Building or project layout and size, weather, material and equipment availability, lead craftsmen and foremen availability, subcontractor capabilities, city work hours and work day restraints, difficulty of work, congestion, and even parking for the craftsmen. The following example provides several schedule duration options based upon an estimated concrete formwork productivity rate of 0.085 man-hours per square foot (MH/SF).

The fire station has 491 feet of continuous footings, three feet wide, two feet deep, that takes 0.085 MH/SF to form.

The total formwork, including both sides is 982 LF, and at two feet deep equals 1,964 SF of forms.

$$491\,LF \times 2\,Sides \times 2'deep \times 0.085\,MH / SF = 167\,MHs$$

167 MHs worth of formwork can be built with a variety of carpenter crew makeups, including:

1 carpenter working 167 hours or ~ four weeks

2 carpenters for ~ two weeks

4 carpenters for ~ five days

10 carpenters for ~ two days

20 carpenters for ~ one day, or

167 carpenters working one hour

Not all six of these scenarios are reasonable; for example, options 1, 2, 20 and 167 can easily be eliminated from consideration. Which of scenarios 4 and 10 is the most reasonable? The answer to that and many other scheduling questions is "it depends." The computer alone cannot make that determination. Even with preprogrammed crew makeups, such as a C4 crew is comprised of one working carpenter foreman, two journeymen carpenters, and one laborer. The superintendent still needs to make a judgment based on the parameters discussed above. Alternate decisions involve the quantity of crews employed, such as one or two or three. Are any of the craftsmen working on preassembly? What form system will be used? What equipment is available? This is where the scheduler must consult with his or her superintendent.

The construction schedule can also be utilized to develop a manpower plan. The process to prepare a manpower curve or histogram is a similar process to creating a cost-loaded schedule and cash flow curve as will be discussed in Chapter 11. Hours from the estimate are placed on construction activities and factored for hours per day or per week and calculated crew sizes. The superintendent will want to avoid large peaks and valleys in manpower, but slight fluctuations are unavoidable. Manpower stacking or "stacking of trades" often occurs toward the end of a project and has a negative impact on labor productivity. A sample project labor requirement curve is shown in Figure 10.1.

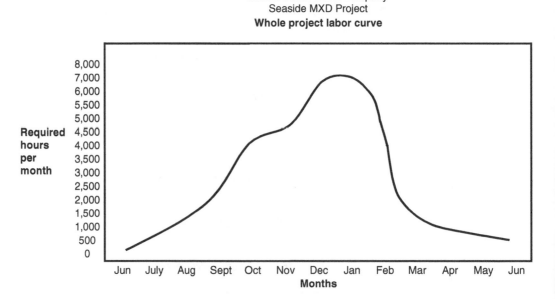

Figure 10.1   Whole project labor curve

Manpower can be increased beyond the superintendent's core group by calling the union hall (for union contractors) or going to labor resource agencies (for merit shop contractors), but neither of these sources provide craftsmen who are necessarily loyal to the contractor, and care must be given to choose carefully for quality and safety and security reasons, among others. Selective overtime (OT) may be awarded to the core team, which in the long run, may be more cost effective than hiring large influxes of crafts off the street. Larger GCs can move people between different projects, which is part of the role of a home office general superintendent. There is always fill-in work on a jobsite and a savvy superintendent will have a pocket-full of these activities around to keep his or her core team busy.

Scheduling manpower requirements utilizing complicated formulas and algorithms is also possible by the use of software, but it is not necessarily practical. Imagine you are the project manager and you provide the following if-then formula to your project superintendent and tell him or her that this is your solution to their manpower balancing challenge. If so, you better have brought your hard hat to work that day!

$$\text{If } (A,F) - D \times (Dw - Dy - Dz) \times (Fy)$$

There are many potential reasons for missing the schedule, beyond incorrect or fluctuating crew size, and it is important for the superintendent to understand that any or all of these may impact his or her chance of success. Potential reasons for missing the schedule include:

- Inclement weather;
- Unforeseen conditions;
- Original schedule error, such as a missed restraint;
- Field (craft or foremen) not performing up to expectations;
- Owner changes;
- Designer impacts;
- Management errors, such as delayed material orders and submittal processing, and others.

Construction Crews

As indicated, there can be many reasons for lack of crew productivity; one additional potential is improper crew size. The superintendent cannot call the union hall on Monday and ask for 20 carpenters, lay 10 of them off on Tuesday, lay another 5 off on Wednesday, call the hall for an additional 10 on Thursday, and request 5 more on Friday. Craftsmen have family and personal obligations and they return steady

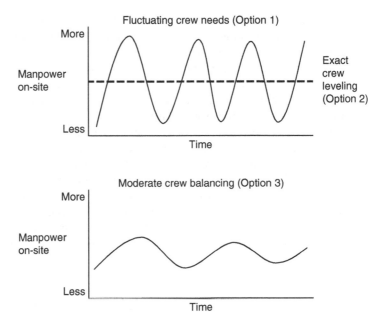

Figure 10.2 Construction crew balancing options

employment from the contractor with loyalty and good performance. But the superintendent cannot keep exactly the same crew and crew size level for the duration of the project either. There will be some necessary fluctuations as shown in Figure 10.2, and a middle ground is likely the most optimal position.

Crew balancing is similar to pouring moderate sized concrete pours. A pour of 20 cubic yards (CY) is not cost effective and not worth mobilizing a concrete pump for but 2,000 CY is likely too large. Concrete finishers will fluctuate in and out of a project, but if their crew size can be moderated, such that they are challenged but not overburdened where safety and quality are affected, they will appreciate the steady work and the superintendent will be rewarded with a consistent good concrete slab finish. This is similar to construction equipment rental efficiency as will be discussed later.

Contractors still use labor union hall terms like "calling up the hall" or "people on the bench," but today most unions do not have a physical bench with several craftsmen waiting to be dispatched to a project; it is all done electronically or through networking. Union craftsmen work on certain scopes. A long list of craftsmen or tradesmen and their responsibilities along with the types of contractors they typically work for is included on the book's companion website. (Let me know if you have others to add to this and I will update the information in the next edition!) There are jurisdictional rules regarding "who works on what" and these are agreed to among the union trades and described in the Green Book, which is more technically known as *The Plan for the Settlement of Jurisdictional Disputes*. Because of this, union contractors are not able to be 100% flexible with moving craftsmen around and balancing their labor needs. For example, if the concrete laborer was done placing concrete in the slab-on-grade by noon, he could not grab a trowel and finish the slab, nor could he begin installing rebar in the wall or build the forms for the columns – this work is all the responsibility of other labor trades. Merit shop labor is generally more flexible about performing a variety of work tasks.

Table 10.1 includes a sampling of crew makeups and costs. A man-hour, in construction terms, is an hour of a craftsmen's time. Crew hours are the amount of hours the crew incurs every hour of the day, or hours of the week the crew works. There are not set sizes of crews; virtually any number and combination of foremen, journeyman, and apprentices is possible, including the mix of craftsman types, such as carpenter versus ironworker. But many superintendents feel that a crew of more than eight craftsmen becomes unmanageable. Very small crews, such as three, will likely include a 'working foreman' or one journeyman is assigned the role of 'lead' and the crew is without a foreman. Some estimating databases, such as *RS Means*, include typical crew makeups.

A learning curve reflects that the crew will typically start out slow on day one, improve productivity on day two, and by day three they are producing at or beyond the productivity rate the scheduler and estimator had planned. It is similar to getting out of the car and rushing to the first tee on the golf course and hitting your drive in the state championship tournament, without hitting a bucket of balls or even taking a practice swing. Learning curves incorporate real cost and time activities, such as shake-out, layout, tool-up, unload, stage, prefabrication, and others.

Table 10.1 Construction crew costs

Crew	Predominant Craft (2)	Quantity	Predominant Craft wage Per/HR	Crew Hours/day	Crew Cost Per/day	Crew Hours/WK	Crew Cost Per/WK
C4	Carpenter	4	42	32	$1,344	160	$6,720
CM3	C. Mason	3	45	24	$1,080	120	$5,400
IW7	Ironworker Rebar install	7	44	56	$2,464	280	$12,320
IW16	Ironworker Struct steel	16	44	128	$5,632	640	$28,160
L3	Laborer	3	36	24	$864	120	$4,320
L8	Laborer	8	36	64	$2,304	320	$11,520
OE1	Operator	1	41	8	$328	40	$1,640
T1	Teamster	1	40	8	$320	40	$1,600

Notes:

1: Many crews include a mix of craftsmen, such as C4 includes three carpenters and one laborer and crew IW7 includes five ironworkers (including one working foreman) and two equipment operators

2: Labor burden would need to be added to all of these hourly, daily, and weekly labor costs

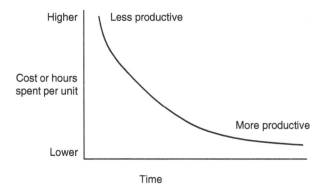

Figure 10.3 Productivity versus time trade-offs

Similar impacts to productivity result from schedule interruptions. If the crew reaches its optimum rate, but then is moved to another task due to a lack of material or equipment or changed conditions, productivity is impacted and momentum is lost. If the crew were sent home for a few days due to an interruption, the crew may be lost as well and moved on to another steadier, more secure project. Figure 10.3 reflects productivity improvement over time. Any interruption causes the curve to be shifted up and to the right and somewhat re-created. Work interruptions may be a source of contractor claims, as discussed in Chapter 18.

Overtime Affects

Overtime can pay off and be productive, to a point, but then after that it is counter-productive; efficiency goes down and costs go up. It is difficult to determine exactly where the optimum point is. Schedulers have complicated OT output formulas stemming from production industries but these are difficult to apply to construction. All agree that there are productivity impacts with working extended OT; it is just difficult to prove exactly how much. Most claims consultants simply use an as-planned cost estimate and schedule and compare to an as-built cost estimate and schedule, which does not consider the contractor's own inefficiencies or estimating and scheduling errors. The difference between the two is the amount claimed as shown in the next formula.

Actual cost − Estimated cost = Cost overrun

Cost overrun + Impact costs including general conditions and fees = Amount of claim

Table 10.2 Overtime labor productivity

Days per week	Hours per day	Hours per week	Efficiency
4	10	40	100%
5	8	40	100%
5	9	45	89%
5	10	50	80%
5	12	60	67%
6	8	48	83%
6	9	54	74%
6	10	60	67%
6	12	72	56%
7	8	56	71%
7	9	63	63%
7	10	70	57%
7	12	84	48%

The exact impact to productivity by working extended overtime is virtually impossible to prove as the exact same circumstances would need to be created to compare straight time versus overtime. This comparison would require: The same size of crew and the same crew members performing exactly the same work on the same day with the same weather and other conditions. There are many studies of the impacts of OT on productivity. Table 10.2 includes some comparisons, and the next set of equations combines factors from both Tables 10.1 and 10.2 to compare the cost for the contractor to have a carpenter work five 10-hour (HR) days per week (WK), or 50 hours per week, in lieu of 40 hours per week. In this one example the contractor had to pay for 15 hours of work to realize an addition of four hours of value. There are many other comparisons that could be made, a couple of which are included as end-of-chapter exercises.

$$\text{Straight time} = 40\,\text{HRs} / \text{WK} \times \$42 / \text{HR} = \$1{,}680$$

$$50\,\text{HRs} / \text{WK} = 40\,\text{HRs straight time}\,(\$1{,}680) + 10\,\text{HRs OT}$$

$$10\,\text{HRs OT} \times \$42 / \text{HR} \times 1.5\,\text{OT premium} = \$630$$

$$\text{Total cost of } 50\,\text{HRs} = \$1{,}680 + \$630 = \$2{,}310$$

Efficiency of 5 − 10 HR days / WK = 80%

Total value of work gained = $2,310 x 80% = $1,848

Value gained of 50 versus 40 hours = $1,848 − $1,680 = $168

Equivalent hours gained for 10 HRs OT = $168 / $42 / HR = 4 hours

Cost of value gained = $2,310 − $1,680 = $630

Cost per hour gained = $630 / 4 hours = $157 / hour

There are additional studies that chart the long-term effect of overtime. These studies show that efficiency ratios in Table 10.2 are reduced even further when the craftsmen have worked their third or fourth week of scheduled OT. Some of these reflect that the efficiency drops an additional 5–10% in week three and more after. This all supports the conclusion that OT should be utilized only when necessary and it is best to use it sparingly or unscheduled and standard workweeks should be alternated in between overtime periods.

Overtime impacts are difficult to compare/measure/prove, but all agree that productivity has to be impacted when extended or planned overtime is worked. An occasional 9- or 10-hour day or one Saturday a month often is completely productive, especially if the extra work allowed a follow-on activity to commence the next business day. For example, keeping two laborers two extra hours after a slab pour to place curing compound allows the wall layout crew to begin work first thing the next morning.

Many superintendents reward only certain craftsmen and crews and subcontractors with overtime, but not the whole project. Overtime should not be expected; the less it is planned the more productive it will become. For example, if the crew knows on Monday they will be working the coming Saturday there is a theory they will pace themselves such that the employer still gets only 40 hours of productive work but paid for six days. Whereas if on Friday morning the foreman offers OT to a select few who worked well the entire week, Saturday will also be productive work. Overtime should not be owed to the craftsmen; it should be earned. The impacts of working OT and impacts to productivity are highlighted in the next two examples and another example is discussed in Chapter 18. Some craftsmen choose which projects they work on based upon OT potential.

Example 10.1

When construction is booming, contractors often need to guarantee overtime to attract craftsmen. This was especially common in the late 1980s when there was a large demand for pipe fitters and ironworkers, especially welders. The crafts would travel from project to project, depending on which ones would guarantee them the most OT, all the while living out of travel-trailers. These pipe fitters and ironworkers were known as "boomers" because they were following the building boom. Once the OT was cut back, they packed up and would move on to the next project.

Example 10.2

Overtime, if worked selectively, can be very cost effective. Unfortunately, some projects will not work specific subcontractors or craftsmen on select OT for specific tasks, but require the entire project to work OT. Some superintendents also feel they need to reward all of the craftsmen the extra time and money for fear of reprisal from those who are working only 40 hours a week. This scheduler would come in early Monday morning to review weekend time cards from the 1,000 electricians on the project, and inspect the field work, and report to management on the noncritical activities that were worked on Saturday and Sunday for a premium. It was not a pleasant task and the electrical superintendent did not appreciate the oversight.

An alternative to working extended periods of overtime is to simply increase the crew size, say from five carpenters on crew C5 to eight. This is also not as simple as it sounds. First the scheduler is assuming that there are eight qualified and dedicated craftsmen available to work on the task. Second, it assumes that there is enough room for them to work alongside the crews that are already on the project. Physical space in a project is a limited resource, as well. In addition, the extra carpenters would need to have sufficient materials and equipment and qualified supervision; these are also limited resources. The concept of crew stacking, or stacking of trades occurs when contractors choose manpower increases on a straight-time basis over OT for the crew that is already mobilized. Studies reflect that a crew size increase of 10% may lose 2% productivity, 50% increase may lose 7% productivity, and 100% increase in manpower may lose 15% productivity or more. These inefficiencies are assuming that there are sufficient materials, equipment, supervision, and physical space to work in. Similar to OT impacts discussed above, it is reasonable to assume some impact occurs but it is difficult to quantify the exact effect of increased crew size.

10.5 INDIRECT RESOURCES

Indirect construction costs are those that are not readily visible during or after construction but are still necessary. These include the administrative cost both at the jobsite and in the home office, along with construction equipment use. Indirect costs are also known as overhead, general conditions, or general requirements. These administrative efforts are also limited resources that the jobsite team relies on.

Jobsite General Conditions

Jobsite indirect labor includes the project manager and superintendent. Many contractors will boast that their most valuable asset, or resource, is their people, especially these two categories of people. One simple resource balancing tool for the whole company is through the use of their collective summary schedules. Each project is simply represented by a single line on the bar chart as reflected in Figure 10.4, which was prepared on Microsoft Project. Each project bar is backed up with a summary schedule that is included in the PM's monthly executive report or monthly cost and fee forecast discussed elsewhere. Each of those summary schedules is in turn backed up with a detailed schedule posted to the conference room wall at the jobsite.

In addition to these summary bars representing individual projects, they represent how the company's resources are being used, in this case the PM and superintendent. After a project is finished, those two resources are freed up to begin another project. The company does not have unlimited resources in any category, especially with respect to experienced PMs and superintendents. Yes, additional people can be hired externally or groomed internally, but it is a long and sometimes risky process. This corporate schedule also reflects the company's backlog – how much work is underway and what is coming up.

Construction Equipment

Having too many resources available on a jobsite, such as construction equipment, or even in the company's storage yard, does not connect with lean planning techniques as presented in Chapter 5. Every superintendent would like to have extra ladders, scaffold, welding machines, compressors, form lumber, scissor lifts, and other tools safely stored on the jobsite just in case he or she might need it. Equipment discussed in this book generally refers to construction equipment utilized by contractors to build the project. Other types of equipment that are a permanent part of the building include mechanical air handlers, electrical generators, and the project owner's process equipment.

The easiest method to manage equipment rental or lease from a project manager's or superintendent's perspective is from outside third-party suppliers that own and rent construction equipment as a main source of business. This is compared to renting from an internally owned equipment company. Third-party equipment is expected to be delivered to the jobsite in perfect condition and is expected to remain that way

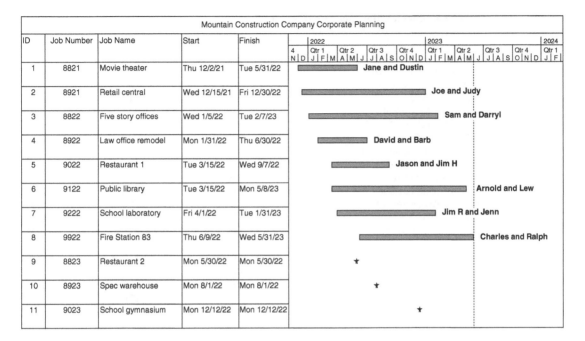

Figure 10.4 Corporate summary schedules

during the course of construction. Equipment breakdowns have a significant effect on jobsite labor productivity. Contractors can obtain bids from outside sources and negotiate rental rates and terms and conditions. Conversely, the jobsite management team cannot really negotiate with their own construction company on the rental of one of their internally owned forklifts.

A very important aspect to equipment rental is the management of economical rental durations. This is easier to control with outside rental companies than with internally owned equipment. Equipment rental often is accompanied with mobilization or delivery charges and demobilization or pickup charges. These charges can be substantial and repeated rotation of the same piece of equipment on and off the jobsite can add up. The project superintendent should keep a watchful eye on this to ensure efficient equipment use. There are also more economical durations to rent equipment, e.g. it is more economical and potentially less expensive to rent an air compressor for:

- One whole day rather than six hours,
- For one whole week rather than four days,
- For one whole month rather than three weeks, and
- For one whole year rather than 10 months.

Table 10.3 Construction equipment rental rates

Description	Daily	Weekly	Monthly
Air compressor, 185 CFM DSL	$120	$460	$1,085
Crane, truck-mounted, 14-16T	500	1,760	5,196
Vibration plate, medium	75	290	660
Dump truck, 5YD	330	1,310	3,260
Water truck, 2,500 Gal	335	1,310	3,910
Forklift, 40'-42', 6,000#, 4WD	290	1,100	2,600
Forklift, 50', 8,000#, 4WD	360	1,320	3,960
Forklift, 55', 10,000#, 4WD	430	1,625	4,630
Generator, 45KVA, 32KW	180	610	1,910
Loader, skid, medium	195	715	1,760
Excavator, mini, 7,500#	245	915	2,590
Backhoe, 4WD, extendahoe	270	1,020	2,910
Boomlift, 40'-42', ST 4WD	355	975	2,310
Boomlift, 60', ST 4WD	440	1,225	3,250
Boomlift, 80' w/5' Jib, 4X4	867	3,300	7,860
Scissor lift, 19' vertical	115	310	770
Scissor lift, 26' vertical	115	430	1,010
Scissor lift, 32' narrow	145	535	1,260
Trash pump, 3"	75	195	555
Shopvac, wet/dry	35	95	240
Welder, 225 Amp, gas	55	160	475

A very good local Pacific Northwest equipment supplier's partial rental rate sheet is included here as Table 10.3. It is easy to see the economics of managing rental periods from this table. An equipment schedule is utilized to track these rental periods and is a valuable control tool for contractors. An equipment schedule example is included in Chapter 14 with other scheduling tools.

Home Office Resources

Some home office resources can be apportioned to different projects (say one day a week or full time if necessary), based upon the project's particular situation, size, complexity, and contract conditions. This is also an example of activity-based costing; charging costs to actual work activities whenever possible. Home office resource candidates include some of the following:

- Specialty superintendents, such as an ironworker superintendent, earthwork superintendent, or concrete finisher superintendent (see the next example);
- Quality control officer or inspector;
- Safety officer or inspector;
- Cost engineer;
- Staff scheduler, and others.

Example 10.3

This large general contractor had several specialty superintendents. These were mostly senior superintendents who knew their trade well, such as steel erection or concrete finishing, but were not necessarily equipped to deal with other subcontractors, the project owner, or city inspectors. They were respected by the GC's project superintendents and called on as needed on a project-by-project basis. One particular specialty superintendent was very familiar with the work of heavy civil subcontractors. He had come up through the field as an equipment operator and foreman and was familiar with equipment operation and productivity. All of the earthwork and site utility subcontractors knew and respected him as well. He would typically visit several projects a week and his prorated wages (a form of ABC) were well worth the understanding and communication that he brought to each of the GC's superintendents for this very critical schedule aspect of most construction projects.

Contractors also have other corporate restraints on resources that may prohibit them from bidding on a certain type and size of project. Two of these include liability insurance and bonding capacity. A contractor's access to ready cash to pay its short-term project expenditures via a line of credit with its bank will also limit its ability to accept new work.

10.6   REPORTING

Resources can be measured against material quantities, such as cubic yards of concrete, tons of gravel, square yards or tons of asphalt, thousand board feet of lumber, and tons of structural steel. Heavy civil projects often employ quantity surveyors to measure quantities of work physically in place. Percentages complete of individual activities can be calculated through formulas, such as the following.

$$\text{Percent complete} = \text{Cost}(\$) \text{ or hours spent / Total estimate, or}$$

$$\text{Percent complete} = \text{Quantity complete / Total quantity estimated, or}$$

$$\text{Percent complete} = \text{Time elapsed / Total duration scheduled}$$

Other manual methods include the use of material tracking logs, such as a concrete log that tracks the total amount of concrete expected to be placed on the job compared to how much has been delivered and placed. If 200 CY of concrete has been cast of Fire Station 83's total 300 CY, then concrete is approximately 67% complete. Another method is to count dump trucks and calculate the amount of earth imported or exported compared to the total estimate.

An unscientific measure of progress is simply to ask the superintendent what percent complete is the project? Some may feel this method is subjective and an inaccurate means of determining completion status. This author has had decades of positive experience working with construction superintendents – without them there would not be any construction projects. They may be off by a few percentage points when reporting progress, but not intentionally.

Just as the crane and the crane operator are important tools, they do not run the project, the superintendent does. The scheduler and computer programs that created the schedule are important tools, but they also do not run the construction project. There are many reports that serve as valuable scheduling tools and these are discussed more fully in Chapter 14.

10.7 SUMMARY

Resources are limited for contractors and must be factored when estimating and scheduling construction projects. Particular emphasis is given to manpower, equipment, and cash flow resource management. Resources can be allocated to divisions within a company, projects, and ultimately direct work construction activities; this is a premise of ABC, which supports lean construction planning. Balancing these limited resources is an efficiency goal of any contractor, but forcing a project into a perfectly level resource utilization plan is not realistic and would jeopardize the construction schedule.

Many factors can affect a project's productivity, specifically labor productivity. It is the superintendent and his or her foremen's responsibility to make sure that they have chosen the best craftsmen and provided them with the proper materials and equipment to allow them to be successful. The superintendent's participation in the original schedule development and choice of best-value subcontractors also enhances the contractor's ability to meet productivity goals and return a fair profit for the company.

Overtime can be beneficial to a project if worked selectively. Long-term use of OT is expensive, inefficient, and not healthy for the craftsmen. Increasing manpower in lieu of OT is an option, but it may also have a detrimental effect on productivity. Craftsmen need space to work, tools, equipment, and qualified supervision, and these are all limited resources. Valuable indirect resources include the contractor's PM and superintendent and construction equipment.

10.8 REVIEW QUESTIONS

1. Which should take precedence, a schedule driven by logic and relationships or one prioritized by resource leveling?

2. Which construction craft installs concrete reinforcement steel and structural steel?

3. Why is it not practical for a superintendent to start out day one with a 10-person crew and keep that exact same crew makeup and size throughout the project until final completion?

4. Arrange these schedule development activities in order: (A) Direct labor estimate, (B) Work breakdown structure, (C) Activity durations, (D) Crew determinations, (E) Relationships and restraints, (F) Activity list, (G) Logic diagram, (H) General conditions estimate, and (I) Summary schedule draft

5. On paper, which crew balancing option in Figure 10.2 (1, 2, or 3) is the least expensive?

6. Utilizing equipment rental rates Table 10.3 what would be the cost advantage of renting a truck crane for one month rather than 17 days or 3.5 weeks?

7. Were any home office costs included in the case study's jobsite general conditions located on the book's companion website?

8. Other than the GC's project superintendent saying his or her project is approximately 30% complete, how might they back that up or be more definitive?

9. What is wrong with calculating the cost of an impact claim by simply subtracting the estimated cost (or schedule) of the work from the actual cost (or schedule) incurred?

10.9 EXERCISES

1. Assume resources are limited and address the following 'what if' scenarios. (A) What if a contractor needed a small crane but there was not one available? (B) What if a 1,000 SF slab-on-grade required to be poured (and finished) on Friday, but the contractor's only finisher was ill? (C) What would happen if the project owner suffered financial difficulties on another project and the bank was restricting its loan on your project? (D) What would happen if the plumbing subcontractor pulled half of its crew off of your project in favor of another?

2. Utilizing the productivity table in this chapter, what percentage return on productivity, would a contractor achieve by working seven 10-hour days? Assume 100% productivity is achieved by working five 8-hour days a week.

3. Assuming the craftsmen in Exercise 2 earned $50 per hour base wage, and 150% for overtime, and 200% for Sunday work, how much would a craftsman cost his or her employer for that week?

4. Combine Exercises 2 and 3 and calculate the productive wage of the craftsman.

5. Other than the home office resources listed in this chapter, what other home office expense may be a good candidate for ABC?

6. Who on Mountain Construction's team should be targeted for upcoming projects 8823, 8923, and 9023 in Figure 10.4, and why them?

7. Why might an equipment rental company offer economy of scale rental rates, such as a week being less expensive than four days?

8. Combine the fire station formwork crew analysis from this chapter with the detailed schedule included on the book's companion website. What crew size (C4 or C10) did the superintendent choose? You will have to do some interpreting for this exercise.

9. How long would it take a C7 crew to construct the fire station continuous footing formwork? How long would it take two C4 crews to build the formwork?

10. What was meant by the phrase "qualified and dedicated craftsmen" regarding increasing crew size? Aren't they all qualified and dedicated?

C h a p t e r **11**

Cash Flow Schedule

11.1 INTRODUCTION

There are a variety of tools that are used in construction. Architects use computers today, not drafting boards, and produce drawings with software, such as computer-aided design and building information modeling. Surveyors used to use a level or transit but today they most likely use a laser or total-station. Carpenters use hammers and plumbers use pipe wrenches in the field. Project managers, superintendents, and project engineers use a variety of construction management and scheduling tools as discussed throughout this book, most of them prepared and transmitted on the computer as well.

Even though many construction management tools have changed, there is one tool that project owners have used with general contractors (GCs) and GCs likewise with subcontractors for hundreds of years, and will continue to do so into the future, and that is *cash*. Project owners also have the next potential contract as a tool to use with GCs, and GCs do the same with subcontractors, but without cash, contractors will experience financial difficulties and potentially suffer bankruptcy. Cash for this discussion is not necessarily actual dollar bills and coins but a positive flow of money through the bank. In fact, very few contractors will deal in hard currency and those that do may be trying to avoid various taxes. Even contractors who report profits, have a good reputation for quality work performed safely, and bring their projects in on time, may still have financial difficulties if they are not being paid on time and do not have a positive cash flow.

The lack of a good positive flow of cash has an even greater effect on subcontractors and suppliers, especially those that are subcontractors to subcontractors and are far removed from the client and the bank. An organization chart included in Chapter 17 depicts a subcontractor's distance from the client. The lack of cash flow is likely the single most common reason for contractor failures.

The focus of this book has been on construction planning and scheduling efforts of the jobsite team, and as stated throughout, "time is money." The process of preparing a construction schedule, and implementation of that schedule on the jobsite, has a direct impact on cash control. Cash flow is an important focus of the contractor's chief financial officer (CFO) and chief executive officer, along with other stakeholders including boards of directors and equity partners. Each jobsite is an individual revenue base; that is one of the differences between construction and other industries. The home office relies on its construction projects to bring in cash in the form of monthly payments from their clients and add together all the project revenues and expenditures to support corporate cash flow needs. Many CFOs will use a positive cash flow generated by the construction teams to produce other income in the form of short-term investments, such as stocks and bonds, equipment company operations, and real estate investments. The home office cash flow position is beyond the responsibility of an individual project manager (PM) and superintendent, but each of their jobsite cash flow efforts contribute to the corporate bottom line.

This chapter will discuss in detail the process of preparing cash flow schedules or curves, which requires first the creation of the project estimate and schedule. These two documents combined then create a cost-loaded schedule. The concept of cash flow has several elements; most of them included in the broad concept of cash outflow and cash inflow. All the different jobsite expenditures including labor, material, equipment, subcontractors, and indirect costs have a negative impact on cash flow and require tracking. The only significant positive flow of cash for GCs is revenue received from the client. Contractors always want to operate in the black so that their inflow of cash is greater than their outflow. There are various methods a contractor can use to improve its cash position and those are discussed here as well. Some of them are ethical and some of them are not.

11.2 CASH FLOW SCHEDULE PROCESS

A cash flow curve is a projection of the total value of the work to be completed each month during the construction of the project. It is created by cost loading the construction schedule and plotting the total monthly costs as reflected in the next equation. Often, it is one of the first things the project owner will ask of the GC and may be required by the construction contract. The fire station's special conditions to the contract, section 013200.2.3 regarding cash flow states in part: "Submit a diagram within 14 days of issuance of a notice to proceed . . . including a cash requirement prediction based on indicated activities . . ." One reason this is required is to provide information to the lender for anticipated monthly payments. Some PMs resist on the basis that the cash flow schedule or curve will be wrong and that they may be penalized for it. The most important thing the GC's PM does is to get paid from the owner for the work that has been completed on the project. This will be discussed in more detail in "Jobsite Control Systems" in Chapter 15. If a cash flow curve is a requirement to facilitate payment, it should be developed.

The cash flow curve is easy to prepare and begins with the development of a cost-loaded schedule. Development of the cost-loaded schedule by the estimator or PM

Estimate + Schedule = Cash flow schedule

Cost-loaded schedule

starts with a summary schedule and a summary estimate. Detailed versions of these may be helpful to prepare the cost-loaded schedule but schedules and estimates that are overly detailed with hundreds or thousands of line items might be cumbersome. Schedules and summary estimates with less than 25 activities are probably too few and 40–50 activities would be ideal. Anything with over 100 would still be usable, but potentially unnecessary. The list and description of the activities on both the summary schedule and the summary estimate should more or less be similar. The best method to explain development of a cost-loaded schedule is with a set of step-by-step instructions, as follows.

1. Start with an Excel spreadsheet. List direct work activities down the left side of the sheet. Add the cost of those activities in the next column. Add a subtotal row below the direct work activities and add the costs vertically down. Verify that this subtotal cost matches the subtotal cost from the summary estimate. Some of this may be cut and pasted from the summary schedule or pay request schedule of values (SOV).

2. Across the top of the Excel spreadsheet list the months from the construction schedule or weeks for a short duration project.

3. Take the estimated costs for each direct work activity and spread them according to when the activities will be complete. Following are four examples of how direct costs might be spread:

    a. If the site utilities are worth $50,000 and will be spent in June, then put $50,000 in June adjacent to the utilities line.

    b. Structural steel installation, estimated at $222,570, occurs in months four and five and can be split evenly at $111,285 for each month. Miscellaneous steel is split out in later months.

    c. Roofing is worth $255,680 and will start the first of October and work continues through November and overlaps into the third week of December. It is suggested that the costs be pro-rated approximately as:

$$35\% - 35\% - 30\%$$

This approximately equates to $89,488 for October and November each, and the last $76,704 in December. Exact estimates and dates are not necessary as this is not an exact science. It may be easier and just as accurate, at the end of the day, to round all these percentages and dollars to the next whole digit. Do not use cents in any of these calculations.

    d. Ten percent of the $200,000 plumbing subcontract is attributed to months one and two for under-slab rough-in, 40% occurs over a two-month span when wall and ceiling rough-in are scheduled, and the balance during the last couple of months of the project for trim and testing. This proportions out as:

$$\$10,000 - \$10,000 - \$40,000 - \$40,000 - \$50,000 - \$50,000$$

If there is a slight adjustment to be made for any of the work line items, do so in the last month. For example, if the plumbing scope was actually contracted for $201,000, place the additional $1,000 in the last month, which would then total $51,000.

4. Total the spread of direct work items down for each month. Labor burden may be split out separately on projects that have a large amount of direct labor, which often occurs early in the project. Labor burden may also be spread proportional to total direct costs, as discussed in step 8.

5. Add a row at the bottom of the sheet below the direct work subtotal row for jobsite general conditions. The general conditions may be spread or proportioned in one of three different fashions:

   a. Evenly spread the general conditions the same amount for each month.

   b. Proportionately spread the general conditions, such that if the general conditions amount to approximately 10% of the total direct work estimate, attribute 10% of the subtotal direct work tally for general conditions across the page. Again, a manual adjustment will need to be made in the last month.

   Options a and b are both easy to compute and easy for a project owner to understand and accept, but neither will be completely accurate.

   c. Calculate approximately how much of the jobsite general conditions will be spent for each month. Most projects realize more general conditions at the front end and the back end of the project, due to activities like mobilization, buyout, and close-out, and a more even spread during the middle of the project. This is subjective and difficult to forecast accurately, as well.

6. There cannot be too many sets of subtotals to keep the costs straight. Add a column on the far-right-hand side of the schedule and add each of the direct work activities across the sheet. This set of totals should equal the original estimates from the far-left-hand side of the sheet which were brought forward from the summary estimate. If there is a mistake, correct this now. If slight adjustments are necessary, make them either in the first or the last month that an activity occurs.

7. Add another subtotal row below the spread of general conditions and add the total direct costs to general conditions.

8. Add another row (or more) for markups. All the markups can be grouped together, labor burden, fee, insurance, contingency, taxes, and others. Calculate the percentage all these markups are of the subtotal for direct and indirect costs. Use this percentage to prorate the markups across the sheet to the right. Most project owners will accept a pro rata share of markups that is invoiced the same way this cash flow schedule is being prepared.

9. Add a total row below the markups and add the subtotal direct and indirect costs to the markups. This total in the column on the far-left-hand side of the sheet and that on the far-right-hand side of the sheet should equal the contract total. If they don't exactly, go back and correct the error.

Table 11.1   Cost-loaded schedule

Project:	Satellite Fire Station 83
Location:	Pasco, WA
Architect:	TCA Architecture
Owner:	City of Pasco

Mountain Construction Company
Cost-loaded schedule
Condensed version

Project duration:	11 months
Project value:	$4,761,841
Date:	5/27/2022
Building SF:	10,612
Esimator:	Charles Kent

Div.	Scope	Cost	2022			Dec/Jan	2023		Totals
			Jun/Jul	Aug/Sep	Oct/Nov		Feb/Mar	Apr/May	
2.1	Utilities	$ 50,000	$ 50,000						$ 50,000
2.2	Demolition	$ 44,409	$ 44,409						$ 44,409
3	Concrete - self-perform	$ 197,559	$ 40,000	$ 80,000		$ 26,000	$ 51,559		$ 197,559
5	Metals - self-perform	$ 370,951		$ 110,000	$ 110,000	$ 50,000	$ 100,951		$ 370,951
6.1	Rough carpentry	$ 181,297		$ 100,000	$ 81,297				$ 181,297
6.2	Architectural woodwork	$ 68,414				$ 34,000	$ 34,414		$ 68,414
6.3	Stone & tile	$ 25,380				$ 12,500	$ 12,880		$ 25,380
7.1	Exterior enclosure	$ 35,968			$ 24,000	$ 11,968			$ 35,968
7.2	Insulation	$ 39,845			$ 26,000	$ 13,845			$ 39,845
7.3	Roofing	$ 255,680			$ 180,000	$ 75,680			$ 255,680
7.4	Weatherproofing	$ 3,451			$ 3,451				$ 3,451
7.5	Caulking	$ 21,890			$ 14,600	$ 7,290			$ 21,890
8.1	Openings - self-perform	$ 95,539			$ 30,000	$ 48,000	$ 17,539		$ 95,539
8.2	Glazing	$ 87,393			$ 25,000	$ 44,000	$ 18,393		$ 87,393
8.3	Overhead rollup doors	$ 122,000			$ 61,000	$ 61,000			$ 122,000
9.1	Drywall	$ 62,696				$ 30,000	$ 24,000	$ 8,696	$ 62,696
9.2	Ceiling systems	$ 17,040					$ 13,000	$ 4,040	$ 17,040
9.3	Painting	$ 59,899				$ 20,000	$ 30,000	$ 9,899	$ 59,899
9.4	Athletic flooring	$ 2,503					$ 2,503		$ 2,503
9.5	Base	$ 5,135					$ 5,135		$ 5,135
10.1	Specialties - self-perform	$ 35,058					$ 22,000	$ 13,058	$ 35,058
10.2	Interior signage	$ 3,500					$ 3,500		$ 3,500
11	Equipment (OFCI)	$ 1,326					$ 1,326		$ 1,326
12	Furnishings	$ 9,047					$ 6,000	$ 3,047	$ 9,047
21	Fire suppression system	$ 75,000	$ 4,000		$ 40,000		$ 31,000		$ 75,000
22	Plumbing	$ 200,000	$ 20,000		$ 80,000		$ 100,000		$ 200,000
23	HVAC	$ 467,300			$ 150,000	$ 150,000	$ 150,000	$ 17,300	$ 467,300
26	Electrical	$ 825,727	$ 45,000		$ 250,000	$ 250,000	$ 250,000	$ 30,727	$ 825,727
31	Earthwork	$ 135,824	$ 135,824						$ 135,824
32.1	Exterior improvements	$ 205,562				$ 90,000	$ 115,562		$ 205,562
32.2	Landscaping & irrigation	$ 67,874				$ 30,000	$ 37,874		$ 67,874
32.3	Asphalt	$ 65,788					$ 65,788		$ 65,788
32.4	Exterior signage	$ 1,500					$ 750	$ 750	$ 1,500
	Subtotal direct costs	$ 3,840,557	$ 339,233	$ 290,000	$ 1,075,348	$ 954,283	$ 1,094,174	$ 87,517	$ 3,840,557
	Labor burden	$ 186,016	$ 31,003	$ 31,003	$ 31,003	$ 31,003	$ 31,003	$ 31,001	$ 186,016
	Jobsite general conditions	$ 393,365	$ 65,561	$ 65,561	$ 65,561	$ 65,561	$ 65,561	$ 65,560	$ 393,365
	Subtotal direct & GCs	$ 4,419,938	$ 435,797	$ 386,564	$ 1,171,912	$ 1,050,847	$ 1,190,738	$ 184,078	$ 4,419,938
	Fee & % markups	$ 341,903	$ 33,711	$ 29,903	$ 90,653	$ 81,288	$ 92,109	$ 14,240	$ 341,903
	Total monthly cost of WIP	**$ 4,761,841**	$ 469,508	$ 416,467	$ 1,262,566	$ 1,132,135	$ 1,282,847	$ 198,318	**$ 4,761,841**
	Total cumulative cost of WIP		$ 469,508	$ 885,975	$ 2,148,541	$ 3,280,676	$ 4,563,523	$ 4,761,841	**$ 4,761,841**

The cost data that is now summed at the bottom of the schedule for each month should reflect the anticipated monthly project expenses. Most of the scheduling software programs can prepare an exact schedule of values with input of the detailed estimate, but again the line items must be exactly coordinated. The computer will not do the logical thinking associated with spreading the estimated costs, but rather just perform the math. The likelihood of the GC being billed by each material supplier and subcontractor according to any anticipated schedule is somewhat remote. The contractor would not normally provide the project owner or the bank all the detail on this cost-loaded schedule; rather they would list just the monthly total figures from the bottom row. These monthly totals may be adjusted for retention and sales tax for pay requests. This then represents the anticipated cash flow schedule, and if producing a curve is not a requirement, the PM may be finished with this effort at this point. The detailed cost-loaded schedule for Fire Station 83 is included on the book's companion website; Table 11.1 reflects a condensed version.

Cash Flow Curve

Now a cash flow curve can be simply plotted. Again, this may be as simple as a keystroke or two with scheduling software. The cash flow curve is displayed with either a bell shape or S shape. The bell-shaped curve represents a plot of the estimated value of work to be completed each month. The S-shaped curve represents a plot of the cumulative value of work completed each month. The best solution is to plot both curves on the same sheet but provide different vertical scales; otherwise the monthly curve is quite flat and does not accurately communicate the change in forecasted cash needs. Some PMs will adjust the monthly figures to reflect a standard or perfect bell-shaped curve. Within reason, this is acceptable for presentation purposes but not a requirement. Many cash flow curves actually depict more of a double-humped camel than a bell. This is caused when there are significant project costs early on in the project, such as prepayment for long-lead equipment and site and structural work, and significant costs late in the project, such as expensive finishes and mechanical and electrical trim. The actual cash flow can later be tracked against this schedule. Figure 11.1 shows the contractor's work-in-place cash flow curve derived from the Table 11.1 cost-loaded schedule.

One interesting twist to the cash flow analysis is that there are several different means of plotting and measurement:

- Committed costs: The purchase orders (POs) have been issued and the subcontracts have been awarded. Therefore, the GC and project owner have committed to spend the money, but it has not yet been paid. This curve is shifted far left of the one drawn in Figure 11.1.

- On-site materials: The reinforcement steel was delivered to the site, but the contractor has not received an invoice for it, and therefore, not made payment.

- Costs in place: The light fixtures that were delivered last month are installed this month. Costs for materials are not usually accounted for until the materials are installed on the project, especially if installed by a subcontractor.

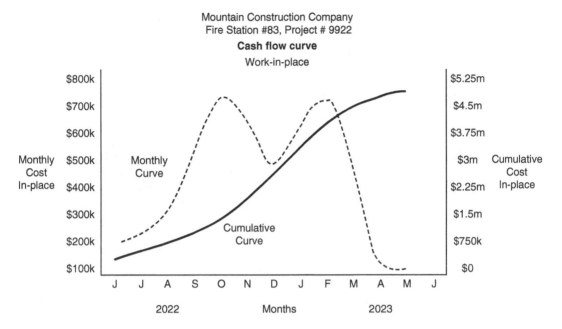

Figure 11.1 **Cash flow curve**

- Costs billed to the general contractor: This reflects invoices received from suppliers and subcontractors. It usually lags and is plotted to the right of costs scheduled to be in place.

- Monthly pay request to the project owner: This follows receipt of invoices from suppliers and subcontractors. The payment request can be up to 30 days after some of the labor was paid and materials were received.

- Payment received: This reflects when payment was received by the GC from the project owner and the owner from the bank. It will lag the formal monthly payment request by 10 to 30 days.

- Payment distributed: This curve reflects payments distributed to subcontractors and suppliers. It will generally lag payment received from the owner by an additional 10 to 30 days. This curve is shifted far right of the one drawn in Figure 11.1.

So, which measure of cash flow should the curve represent? In most instances, the GC's project manager will develop the curve based upon his or her schedule of construction that reflects the anticipated costs in place. This was the method used to develop the curve shown in Figure 11.1. The formal invoice will be one to four weeks behind this curve, and the receipt of cash and subsequent disbursement to subcontractors up to a total of two months behind the time when the work was accomplished. In this way, the actual cash flow will fall behind the original projection. The cash flow schedule, once developed and submitted, should be monitored monthly as another check that the initial project cost control and schedule plan is relatively followed. The preparation of a whole project labor curve presented in Chapter 10 follows a similar process to the development of the cash flow curve.

11.3 JOBSITE EXPENDITURES

Tracking job costs is one of the major and most time-consuming aspects of cost control. Much of the efforts in tracking jobsite expenditures during construction will be the responsibility of the project engineer or cost engineer. In order for the cost control process to be effective, costs need to be accurately tracked and reported prior to construction events. This applies to the reporting of time and schedule control as well. The process of buying out subcontractors and suppliers and drafting and executing purchase orders and subcontract agreements is another aspect of cost and schedule control. Information is gleaned from that process by the GC and clauses are inserted into the agreements which help with these controls. Costs are "committed" when contracts are let for materials and subcontractors and/or when materials are received, and labor is performed on the jobsite. These same committed costs are recorded when invoices and timesheets have been received and/or approved by the jobsite team. The costs are actually expended when checks are cut, or electronic deposits are made. Different accounting systems and different contractors may account for costs incurred under any of these scenarios. In this section, each of the activities and steps associated with tracking direct labor, materials, equipment, subcontract costs, and jobsite general conditions are stepped through.

- Direct labor costs: Labor activities occur during the first week of the month and are paid on Friday of the second week.
- Direct material costs: Materials are received onsite throughout the month. Small purchases may be paid 10 days after receipt of an invoice. Larger material purchases receive a long-form PO and are paid on the 30th of the month, or potentially after the GC receives payment from the project owner, which are the same conditions as those that apply to subcontractors.
- Equipment expenditures: The processing of equipment rental invoices will be very similar to that of materials ordered with short-form POs.
- Subcontractor invoices: Subcontractors perform work throughout the month and incur labor and material and equipment costs similar to a general contractor. Subcontractors are required to submit their pay requests to the GC on the 20th of the month which will forecast the amount of work anticipated to be completed through the end of that month. Subcontractors are paid 10 to 30 days after the GC is paid. This is known as "paid when paid" and often places a financial cash flow burden on the subcontractors, especially subcontractors and suppliers second-tier to the GC's subcontractors.
- Jobsite general conditions costs: Jobsite general conditions include indirect materials, equipment rental, and indirect labor expenses. The invoicing of indirect materials and equipment is similar to short-form PO materials and equipment.

Salaried personnel are generally paid twice monthly, and those intervals vary between firms. For example, they will work through the 15th of the month and then are paid a week later for that half-month, on or about the 22nd.

Jobsite Revenue

As discussed, there are several jobsite expenditures that draw down the contractor's cash flow throughout the month; conversely there is only one source of positive cash influx and that is receipt of payment from the project owner. This is usually once monthly, on or about the 10th of the month after the work was performed. In some cases, owners do not pay until the 30th of the month following the month the work was performed. In the case of speculative residential home builders, they receive one check at the sale or close of each home. Cash flow for speculative builders is therefore very erratic and often requires them to rely on a construction loan to make weekly and monthly payments until a home is sold. Custom home builders will receive monthly draws from their clients similar to commercial contractors. Custom home builders and remodeling contractors will sometimes receive a down payment from the bank, or the client, which allows them some opportunity to operate in the black (cash positive) through the course of construction. So, although cash flow for many contractors follows a typical bell shape or S-curve as presented in Figure 11.1, the revenue curve for a typical commercial GC is a stepped curve in that it is flat and then jumps vertically once monthly, as is shown in Figure 11.2.

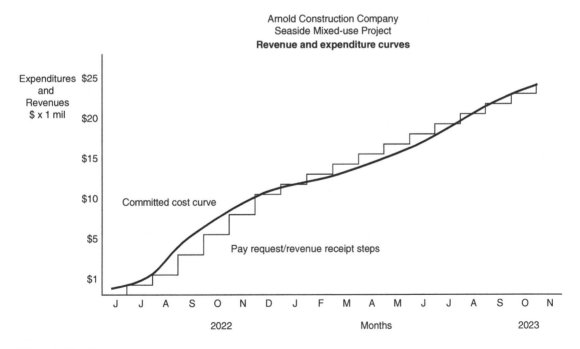

Figure 11.2 Revenue curve

11.4 NET CASH FLOW AND IMPACTS TO HOME OFFICE

The chief financial officer processes cash flow as needed to make payments on all of the contractor's jobsite expenditures and receives monthly checks from all of its clients. The contractor's officers and front-office rely on a positive and not a negative balance of cash flow. Positive cash flow occurs when revenues exceed expenditures. The corporate officers expect more money coming in than going out. The CFO will have an established line of credit with their bank and in the case of a short-term cash short-fall they will call on those funds but pay them back as soon as possible. Unless the construction company is owned by an individual and operates as a sole proprietorship, the company officers are typically accountable to a board of directors and equity investors and partners. These equity partners do not expect to have to dig into their pocket when the contractor is operating in the red (negative cash flow); instead they expect a positive cash flow and an above-market return on equity due to the high risk of owning and investing in construction companies.

Revenues typically lag expenditures early in the project when there is a large amount of direct labor expended to erect the structure. The GC pays direct labor on a weekly basis and does not withhold retention from the craftsmen's checks. Later in the project, such as during enclosure and finishes, there is an increased mix of subcontractors from which retention is being withheld and the GC's cash flow needs are eased. The GC's revenue curve will rise above the expense curve when there are more subcontractors than direct labor and material expenditures. Similar to plotting the cash flow curve, there is no exact date this will occur, but it is usually approximately mid-way through construction for most typical commercial construction projects that employ 80–90% subcontractors.

Methods to Improve Cash Flow

As stated, general contractors are not in the business of providing construction loans to their clients. Speculative home builders will obtain loans and will pay the bank back when they sell the house. These builders do operate in the red and the associated interest costs are factored into their home prices. Real estate developers also obtain construction loans and their loan payments are factored into their pro forma. Commercial GCs expect that their clients will obtain construction loans so that they can keep up with their monthly pay requests; in fact, verification that the owner has financing in place is a good contract execution prerequisite. A long list of jobsite expenditures and the processes necessary to approve invoices and make payments was discussed above, but there is only one source of revenue that occurs monthly and up to 40 or more days after some contractor costs were expended or committed. So how does a contractor operate in the black and not have to borrow money? There are a variety of potentials for the GC to improve its cash flow position at the jobsite level, some of which are *ethical* and legal and are acceptable if negotiated into the contract terms, and others that are very *unethical*. Ethical methods would include subcontracting more work out or convincing the project owner to pay twice monthly. Unethical methods would include falsifying subcontractor invoices or holding excessive retention on subcontractors. A few ethical and unethical examples of affecting cash flow are included in Figure 11.3. There may be serious ramifications for the GC from the client or even from subcontractors if any unethical methods used to improve its cash flow were discovered.

Ethical methods

- Employ more subcontractors and reduce the GC's direct labor
- Include a mobilization charge in the schedule of values
- Client makes a down payment before work commences
- Invoice the client twice monthly
- GC is paid faster, on the 10th instead of the 30th of the month
- Process an early pay request, even for a small amount
- Negotiate a reduction of retention held from 10% to 5%
- No retention held on the GC's labor, material, equipment, or indirect costs
- Release retention early on completed and accepted work
- Beat the estimate and perform work for less cost
- Invoice all insurance and other markups up-front
- Utilize early fee received to offset expenses
- Negotiate a higher fee on direct work than subcontracted work

Unethical methods

- Inflate SOV front-end costs, such as foundations, and reduce finishes costs
- Inflate early SOV activities with disproportional markups and general conditions costs
- Increase subcontractor invoices to the client over their actual billing to the GC
- Hold money back from the subcontractors' checks in addition to the retention held
- Hold the subcontractors' money 30+ days after the GC is paid
- Hold more retention from subcontractors than the client holds from the GC

Figure 11.3   Cash improvement methods

11.5 SUMMARY

Cash is one of the most powerful tools a project owner has with a general contractor and a GC his with its subcontractors. The lack of a positive flow of cash is one of the most common reasons for contractor bankruptcy. Contractors that deliver quality work, safely, on time, and within their budget but must continually rely on the bank to fund its operations will ultimately fail. One significant difference between speculative home builders and real estate developers from commercial contractors is that they obtain construction loans and once the project is completed realize a large influx of positive cash flow. But the interest they pay on the construction loan was factored into their estimates and pro forma. In commercial construction, as with heavy civil and custom home construction, contractors expect that the owner has either obtained the construction loan or has other sources of funding on hand.

Development of a cash flow curve is a simple task and one that the project manager should do to assist the project owner and the bank with preparing for future financing obligations. The cash flow curve is charted from a cost-loaded schedule, which is developed from the contractor's summary schedule and summary estimate. There are several different cash flow curves which can be plotted, but the one that is the most common and straightforward from the contractor's perspective is the work-in-place curve. Processing monthly invoices and receipt of payment will follow a month or so behind this curve.

Net cash flow is defined as revenues less expenditures. Revenue generally occurs once monthly for commercial contractors on the tenth of the month following when the work was performed. Expenses occur at a variety of times throughout the month. Jobsite expenditures include labor, material, equipment, subcontractors, and general conditions. Invoices are processed for all of these expenditures on slightly different tracks. In most cases, payments lag material delivery and installation by a week or two. In the case of subcontractors, they are paid 10 days after the GC is paid by the project owner. This process is known as "pay-when-paid" and helps the GC with management of their positive cash flow. There are several methods that the contractor can use to improve its goal of operating in the black with a positive flow of cash. Some of these are ethical and some of these are unethical. The best method to improve cash flow is to prepare timely, accurate, and fair open-book pay requests.

The next chapter introduces a variety of schedule technology tools that enable the scheduler to produce schedules, prepare cash flow presentations, and assist with implementation of construction management controls. After that, the last part of the book focuses on controls, including schedule control techniques, several jobsite controls including safety, cost, and quality control, and subcontract management.

11.6 REVIEW QUESTIONS

1. How can a cash flow curve be used negatively by a project owner or bank towards a contractor?

2. Why is direct construction work performed with a contractor's own workforce considered riskier and therefore they expect a higher fee?

3. Why should project managers view cash flow curves as a "get-to" and respond gladly when a client requests one be developed?

4. The cost-loaded schedule is an intermediate step in developing a cash flow curve. What are two documents necessary to prepare the cost-loaded schedule?

5. Should a cash flow curve always be a perfect bell-shaped curve? Explain why or why not.

6. There are three different methods how jobsite general conditions are factored into the cash flow projection and are ultimately invoiced. Name them.

7. Describe how the cash flow curve for Mountain Construction Company would change if:

   A. The project start date was delayed for one month.

   B. All of the site improvements were rescheduled to be accomplished during early shell construction rather than near the completion of the project.

   C. The project owner promised the sole-sourced apparatus bay overhead door vendor a 50% down payment upon submittal review. MCC's submittal schedule is included in Chapter 14.

   D. Retention will only be held on the GC's subcontractors and no retention held on direct or indirect costs.

11.7 EXERCISES

1. What would happen to a project manager who (A) operated in red for the first time on his or her construction project, or (B) repeatedly operated in the red on all of his or her projects?

2. Is a GC-subcontractor "pay when paid" contract clause ethical? Is it legal?

3. What would happen to a GC's cash flow if it had been scheduled to receive its invoice from the client on the 30th of month two but did not receive it until the 30th of month three? What would happen to subcontractors of subcontractors if all were subsequently paid 30 days after the GC's receipt of payment?

4. List two additional ethical and legal possibilities a contractor could use to improve its management of jobsite cash flow beyond those discussed here.

5. List two additional unethical methods a contractor may utilize to improve its cash flow position.

6. Prepare an argument for your client why you should receive a 10% down payment before you start any work on their project.

7. Assume a GC has the following expenses for month one of their project. Will they be in the red or black when they receive payment, less 10% retention but including a 5% fee, on the tenth of month two, and by how much?

- Direct labor of $60,000 committed to evenly throughout month.
- Short-form PO direct materials received and invoiced on the 15th for $40,000.
- Long-form PO materials are received and invoiced on the 15th for $70,000.
- $150,000 worth of subcontracted work performed and invoiced on the 20th, which includes a matching 10% retention clause. Subcontractors are paid 10 days after the GC is paid.
- Indirect labor at $50,000 per month payable half on the 22nd of month one and the second half on the seventh of month two.
- Indirect material costing $30,000 and invoiced on the 30th of the month.

8. Assume the same expenditures as Exercise 7 above but now the client does not pay until the 30th of month two. Is the GC now in the red or black and by how much?

9. Assume the same expenditures as Exercise 7 above but now the GC holds the subcontractor and major supplier checks for 30 days after the GC has been paid by the client on the 10th. Is the GC now in the red or black and by how much?

10. Draw a cash flow chart for a third-tier supplier, such as the electrical subcontractor's light fixture supplier. You may need to draw a quick organization chart. Assume the lights are delivered January 15 and installed on February 1. Assume standard contracts and pay periods and processes and predict (A) when is the earliest the supplier might be paid? And (B) when is the latest the supplier might be paid?

Schedule Technology

12.1 INTRODUCTION

Computers are valuable tools to assist in many aspects of construction management including planning, development, and control of the schedule. The identification of the activities, the duration for each, and the sequence in which they are to be completed must be input into the computer for it to develop the schedule. The computer will plot the schedule and calculate the start and finish time for each activity. Computer-generated schedules allow the scheduler and project manager (PM) and superintendent to determine quickly the effects of changing schedule logic, delays in delivery of critical materials, or resource adjustments.

The computer is a valuable scheduling tool, but it cannot take the place of adequate planning. The PM and superintendent should both be actively involved in developing the construction schedule, as they are the individuals who are responsible for completing the project in the desired time. This provides buy-in by jobsite leadership. If they agree with the schedule, they will do everything they can to make it happen. Assigning scheduling responsibilities to a staff scheduler can have its advantages and disadvantages as shown in the next example.

Example 12.1

A civic project worth over $1 billion and scheduled to last almost three years was being built by a national general contractor (GC) with experience in this specific type of work. They had almost 50 people in the jobsite office but still reached outside to hire a full-time scheduler to work on their staff. The scheduler was very adept at the computer side of scheduling, but was not involved in the planning or control efforts of the project. He did not have a hard hat or orange vest or work boots and did not spend any time out in the field. The schedule was several thousand activities long, but was never printed. The contractor did not have the schedule on the wall but relied on the scheduler to pull up portions, or snips of the schedule during meetings and presentations.

Although some old-school builders may still draft their schedules by hand, most are produced by very powerful and versatile scheduling software systems today. As projects become larger and more complex, manual scheduling techniques described previously are not feasible; most projects today have computer-generated schedules. Two of the most popular scheduling programs include Microsoft (MS) Project (Project) and Primavera Project Planner or P6. Many contractors also use Microsoft Excel (Excel) for short-interval schedules. These and other systems are discussed in this chapter along with the advantages and disadvantages of reliance on scheduling software systems.

12.2 SOFTWARE ADVANTAGES

There are many more advantages to scheduling with the computer versus by hand than there is room to list them all. Here are some of the obvious ones.

- One advantage of computer scheduling programs like Microsoft Project and Primavera is they can produce volumes of reports, graphs, histograms, and tables, some of which may appeal to different stakeholders. But these reports must still be a communication tool to be of any value to the scheduler and superintendent. The cash flow curve developed in the last chapter is easily reformatted here into a histogram as Figure 12.1, utilizing Excel.

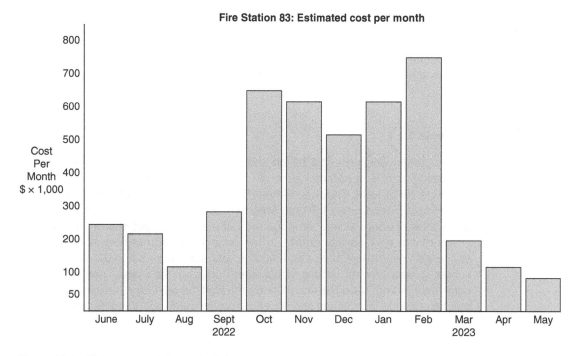

Figure 12.1   Histogram cash flow schedule

- Computer software allows presentation of schedules in many fashions, whether they are simple lists or tables, bar charts, or networks and all can be switched back and forth with a simple keystroke.

- Scheduling software allows the addition of notes to explain lags, interruptions, actual occurrences, manpower spent, deliveries, weather delays, and other valuable information. The scheduler must be careful to avoid information overload. Scheduling programs today provide literally unlimited quantities of relationships, applied resources, reports, et al. Just because the scheduling software can contain almost unlimited data, there is no reason to display it all unless called upon. Overloading the schedule printout with too much data causes congestion – essentially the reader cannot see the forest for the trees.

- Schedule software advantages include the ability to sort. Activities are given codes; there are many complicated ways of coding, some requiring alphanumeric combinations greater than 10 digits and some rely on abbreviations or specification codes, but beyond the coverage here. The computer schedule allows sorting by: GC self-performed work, specialty contractor, building areas (floors or wings), phases, deliveries, milestones, owner or architect activities, construction crafts, and others.

- Another version of a specialty or mini-schedule is a roll-up schedule. The detailed schedule can be "rolled up" to a summary with computer software and compressed. The summary section can later be expanded when that phase of work is coming into view. This is valuable for repeat systems or areas like apartments, shelled office building floors, long stretches of highway or utility projects, and speculative home construction. *Fragnet* is another term for a portion of a larger schedule or network, such as for the roofing subcontractor's work, an example for which is included on the book's website.

- Software automatically performs all of the calculations that are discussed in Chapter 9. This includes immediate determination of all of the following dates and durations without use of a calculator:
  - Early start date,
  - Early finish date,
  - Overall project duration in work days,
  - Late start date,
  - Late finish date,
  - Total float,
  - Free float, and
  - Identification of the critical path(s) of activities.

- Microsoft Project, Primavera P6, and other systems can integrate with project management software systems, such as Procore. This allows the linkage of lead times and procurement activities to the submittal schedule and other scheduling tools, which will be introduced in Chapter 14.

- Colors can be shown in a computer schedule and help communicate the plan by separating work areas or subcontract areas. Color-coded schedules are very readable on the screen, but unless they are also printed in color, which can be expensive if multiple copies are required, the color-coding is lost in black and white prints and copies.

- Personal computers have become very inexpensive and hand-held devices like smartphones and tablets make computer schedule tools accessible for most involved in design and construction.

A few popular scheduling software tools are described in this chapter. Although they have differences and each has its advantages, they have many similar capabilities, including: Spreadsheets, work breakdown codes, calendar views, variety of reports, cash flow curves, earned value, forward and backward passes, summary views, and others.

12.3   MICROSOFT EXCEL

Although not typically thought of as scheduling software Microsoft Excel spreadsheets are used by many contractors for their short-interval scheduling needs. Excel is a valuable scheduling tool and is also utilized for control logs, such as submittal and expediting logs, which will be discussed further on in the book. Excel is also a useful software tool when creating tabular schedules and cost-loaded schedules as has been shown previously. These are easily converted into curves and histograms as seen in Figure 12.1.

Contractors typically develop a short-interval schedule template and share amongst their superintendents so that each project team is not creating their own version. This also provides some uniformity within the company so that stakeholders, such as the foremen, subcontractors, and management personnel are familiar with the schedule. The Microsoft Office Suite allows Excel spreadsheets to be shared in the cloud, which enables project team members to remotely input live updates to the project plan.

The initial three-week schedule for the fire station case study project was prepared by the superintendent using Excel and was included in Chapter 1. Another three-week schedule example is shown in Chapter 7 and the project superintendent's focus on the work happening in early December is included here as Figure 12.2.

Mountain Construction Co.
Short-interval schedule

Project: Fire Station 83, Pasco, WA

Superintendent: Ralph Henry

December, 2022

Date: **Monday, 12/5/22**

Sheet: 1 of 1

No.	Activity description:	S	S	5	6	7	8	9	S	S	12	13	14	15	16	S	S	19	20	21	22	23	S	S	Comments
1	Overhead doors			S	=	=	=	=			=	=	=	=	F										
2	Dry-in milestone														*										
3	Rooftop MEP steel																	S	=	=	=	F			Set equip next week!
4	Install switchgear			S	=	=	=	=			=	=	=	=	F										
5	Energize switchgear														*										
6	Cover inspections																	SF							
7	Insulation																		S	=	=	=			Finish early Jan.
8	Begin hanging interior doors																		S	=	=	=			Finish early Jan.
9	Begin interior glazing systems																		S	=	=	=			Finish early Jan.

Figure 12.2   Microsoft Excel short-interval schedule

12.4   MICROSOFT PROJECT

Microsoft Project was developed in 1984 for DOS use. Three additional DOS versions were released before the first Windows version was released in 1990. Since then about 15 different versions have been released, some within one year of the previous edition. Each version has added features and improvements over the previous ones. The most recent version of Project was released in 2019. A Macintosh version was made available in the early 1990s but has since been dropped.

Many see Microsoft Project as a fairly simple and intuitive scheduling program to learn. It is a popular tool for field superintendents to produce their three-week look-ahead schedules. Most scheduling software programs have the ability to produce both bar chart and precedence diagrams, including critical path networks. Project also features many different tools available for the scheduler, some of these include:

- Project has the ability for schedulers to work in teams.
- Team members can work on the same schedule network remotely and simultaneously.
- Many different views are available, including a calendar view. A calendar schedule will be discussed in the next chapter.
- Project has the ability to schedule and track actual resource usage, such as manpower, equipment, and cash.
- Resource use can be tracked and balanced, or pooled, for multiple projects simultaneously.
- Detailed tasks can be summarized or rolled up into summary tasks.
- Project can handle 10,000 activities and connect with other Microsoft software tools, such as MS Outlook.
- This software can produce a variety of cost and schedule management reports, including graphical reports. These graphs are capable of comparing planned, completed, and remaining schedule tasks.
- Project, as with many other scheduling programs, is offered online and has cloud capability. Microsoft Online provides an online project management web application.

Most MS Project schedules default to a typical Gantt chart and are plotted in a traditional waterfall view. The software includes options like Task Board and Sprint, which allow a large schedule to be broken down into manageable parts, similar to specialty networks discussed throughout this book.

Microsoft customers can purchase the software for unlimited use or can purchase a subscription that involves a monthly fee and is supported by MS software engineers. Microsoft Project is available in two versions, Project Standard or Project Professional. The Professional version has more options and costs about 70% more than the Standard version. The Professional version also has additional team collaboration tools available for the scheduler.

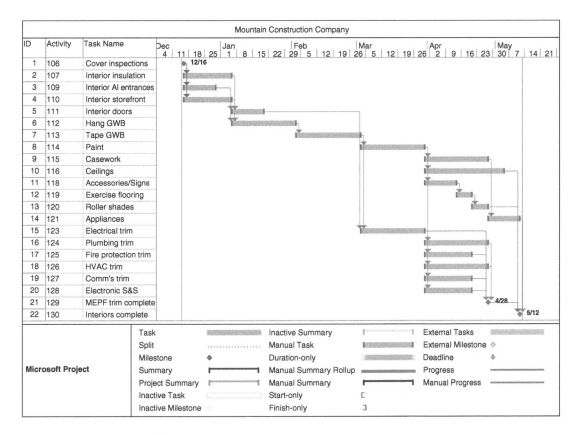

Figure 12.3 Microsoft Project schedule example

Microsoft Project was the required software for bidding general contractors to use in the book's fire station case study project. Many MS Project examples have been utilized throughout this book and others are available on the book's website. Another example is included as Figure 12.3. Additional information about MS Project can be found at https://www.microsoft.com/en-us/microsoft-365/project/project-management-software.

12.5 PRIMAVERA PROJECT PLANNER

Primavera Project Planner P6 is an enterprise project portfolio management software. It includes project management, scheduling, risk analysis, and resource management, and collaborates with other construction management software systems. Primavera was launched in a DOS version in 1983. The Primavera P3 Windows version was released in 1994. Previous versions were known as P3 and Sure-Trak. These versions were utilized substantially by the construction industry, especially heavy civil construction. They were retired in 2010 in favor of P6. Similar to MS Project, there have been several versions or improvements issued. The company was purchased by Oracle Corporation in 2008.

Primavera is generally thought to be a larger and more complicated scheduling system than MS Project. It is often specified by public project owners as the required scheduling system for their projects. P6 features are very similar to MS Project, and include:

- iPhone and web interface,
- Critical path method,
- 100,000 activity projects,
- Lean scheduling concepts,
- Upstream portfolio planning,
- Downstream risk mitigation,
- Resource management,
- 150 reports,
- Portfolio Management allows integration of multiple projects,
- Team applications, and others.

Oracle's Primavera Cloud integrates construction project and portfolio planning and multiple delivery teams to enable planning, resourcing, risk mitigation, scheduling, and project management. An example of P6 is included as Figure 12.4. Primavera Contractor is a scaled-down version that can handle projects with up to 2,000 activities and is available online. Additional information can be found at https://www.oracle.com/industries/construction-engineering/primavera-p6.

12.6 TOUCHPLAN

Touchplan is an emerging electronic version of pull planning. Instead of the last planners physically maneuvering sticky notes on a whiteboard, schedulers can create the pull planning schedule remotely and concurrently. According to Touchplan, over 100 GCs have successfully utilized this scheduling system nationally on hundreds of construction projects. Some of the software attributes include:

- Utilizes the master schedule and identifies and clarifies high-level milestones.
- Utilizes pull planning techniques by scheduling backwards from a clear milestone.
- Identifies constraints that could prevent upcoming work from being accomplished as planned.

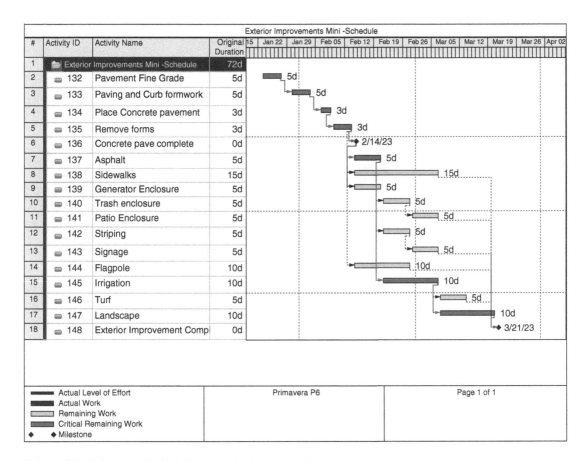

Figure 12.4   Primavera Project Planner schedule example

- It is cloud-based similar to MS Project and P6.
- The schedule is "live" and enables teams to communicate simultaneously.
- Touchplan is accessible anytime with internet connections.

The following quote was provided by a mechanical subcontractor who is using Touchplan for the first time on a downtown skyscraper. An example of a Touchplan pull planning schedule is included as Figure 12.5. Additional information is available at https://www.touchplan.io.

So far Touchplan has been a great tool. All of the foremen are still learning the user interface and all the intricacies but it is getting picked up fast. The key is having all trades interactive, which has been a bit of a struggle with some of the smaller subcontractors.

12/16/2020

Touchplan

Block 24 - Weekly Work Plan

Plans: All Plans Selected
Roles: All Roles Selected
Locations: All Locations Selected (Including Location-less tickets)

3 Weeks Commencing: Monday December 14 2020
Data as of: December 16, 2020 07:37:29 PM EST

Touchplan

Task Description	Prerequisite Work	Project Roles	Prior?	M	T	W	T	F	S	S	M	T	W	T	F	S	S	M	T	W	T
						Week 1 (12/14)						Week 2 (12/21)						Week 3 (12/22)			
Block 24				29	23	44	50	42		4	54	55	66	48			4	68	64	56	65
1–10 Office Space ◆				27	19	42	48	42		4	51	52	61	41			4	60	56	54	59
Floor 3 ◆				17	11	27	37	33			40	44	49	24				40	34	32	36
Trimble I Floor 3		Mechanical	«	4																	
LAYOUT WALLS - INSTALL TOP TRACK I Floor 3		Framing/GWB/ Taping	«																		
Trimble I Floor 3		Plumbing	«																		
OH Rough In-Pre K13 north I Floor 3		Electrical	«	4	4	4	4	4			4	4	4					4			
MEP Core Review and approval I Floor 3		GC - Balfour Beatty	«	1	1	1	1	4				4	4								
Install supports I Floor 3		Plumbing	«	2	2	2	2	2													
Layout Walls I Floor 3		Electrical	«	2	2	2															
Coring Layout I Floor 3		Mechanical	«	2	2	2	2	2			2	2									
Hanger/ South and north I Floor 3		Mechanical	«	2	×	2	2	2			2										
PRE_CON MEP&Framing 1pm on 6th floor I Floor 3		GC - Balfour Beatty			×																
Z-Clip Inspection I Floor 3		GC - Balfour Beatty			×																
OTS PAINT Pre Wall I Floor 3		Paint			×	2															
steel @ grease duct opening (Underslab) I Floor 3		Steel				2															
structural steel @ core holes I Floor 3	Coring Approval Sign off	Steel				2															
FIREPROOFING - FRAMING Z-CLIPS I Floor 3		K-13/Fire Proofing				2	2														
LAYOUT Remaining WALLS - INSTALL TOP TRACK I Floor 3		Framing/GWB/ Taping			4	4	4														

NOTE: *Bold Italic* indicates a constraint. Reason for Variance:
1. Predecessor Activity Incomplete 2. Material Defective 3. Insufficient Labor 4. Coordination / Scheduling 5. Design / Engineering 6. Owner Changes 7. Weather 8. Material Not Available 9. Permit / Inspections 10. Site Coordination 11. GC Directive 12. Finished Before Plan

Figure 12.5 Touchplan schedule example

12.7   OTHER TECHNOLOGY TOOLS

It would be impossible to list all the schedule software programs currently available as new ones are under development constantly; this is a moving target. Improvements and changes are being made to existing systems, adding features and connectivity to other construction management tools, on an ongoing basis. In addition to Excel, MS Project, Primavera, and Touchplan discussed above, this section briefly introduces other systems that are being utilized by our construction industry partners. Many new software systems are being developed, even as of this writing. Some of these systems are also included in this section.

PowerProject by Atlas

Some contractors find PowerProject an easy tool to use out on the jobsite, especially for field superintendents. There are long lists of comparisons between PowerProject and Primavera P6, with PowerProject costing approximately 60% of P6. Both systems have earned value analysis tools. PowerProject is a good tool for smaller projects or subnetworks but may not be as powerful as MS Project or P6, which are popular tools for larger projects. Additional information is available at https://www.projectsanalytics.com/asta-powerproject.

Smartsheet

Smartsheet software captures issues in real time from any device, increases collaboration among project teams, vendors, and clients, and saves time with accurate resource management. It can be integrated with other construction management software systems including Procore, Primavera P6, Microsoft Office, Dropbox, and others. This system reaches beyond scheduling-only aspects and includes quality control, safety control, and budget control aspects of construction management. It has the ability to update actual construction management progress, or lack of progress, easily through the use of mobile devices. Smartsheet connects schedule and construction progress with other departments within the construction company, such as human resources, legal, finance, information technology, and others. Smartsheet also has cloud capabilities. There is a free version with limited access and an expanded paid version. Additional information on this software is available on https://www.smartsheet.com/software/project-scheduling.

Building Information Modeling

Building information modeling or models (BIM) allows integration of various design systems and materials all onto one drawing, or in this case, a computer drawing. The BIM process involves designing a project collaboratively using a coherent system of computer models. The models have transformed conventional two-dimensional drawings into three-dimensional drawings. The architectural, structural, mechanical,

Tuesday 6:22:57 PM 1/12/2021 Day=4 Week=1

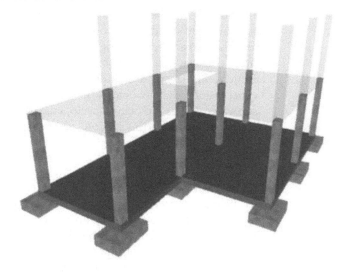

Figure 12.6 Building information model example

electrical, and other designers and systems are brought together to discover potential conflicts before construction occurs. The models can continue to evolve with general contractor and subcontractor input throughout the construction process. The addition of time as a fourth dimension allows the builder to anticipate what the building might look like at any given point in time in the future and a fifth dimension incorporates estimated construction costs. Building information models can provide solutions to multiple sequences that offer different values and options with respect to time, cost, and risk. A four-dimensional BIM model is included as Figure 12.6. Note the schedule date in the upper left-hand corner.

Bluebeam

Bluebeam is utilized by design and construction teams for a variety of functions including estimating, requests for information, as-built drawings, and even pull planning schedules. Figure 12.7 was provided by one of our industry partners as their example of pull planning. This project was being built during the Covid-19 pandemic crisis and the teams were not allowed to all get together and arrange sticky notes, as reflected in photographs in other chapters of this book. Instead they utilized Bluebeam remotely to arrange their schedule activities. Additional information can be found at www.bluebeam.com.

Most of these software systems were not written solely for the construction industry but have been adapted and may or may not fit for a particular section of the industry, such as heavy civil versus residential versus commercial, or for a particular

PULL PLAN: A212D

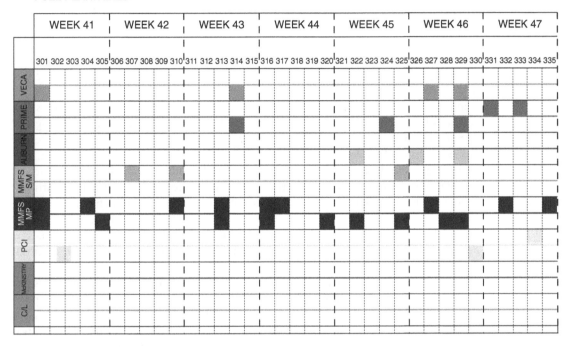

Figure 12.7   Bluebeam pull planning schedule example

contractor. This is not a how-to instruction for any of these software systems, they all have manuals available and their web pages have been provided for the interested reader. There are also many books published on how to use various software programs that facilitates the connection of a particular scheduling system with construction. Many of these software systems also prepare a variety of scheduling reports, some of them automatically. Scheduling reports are discussed along with other tools in Chapter 14.

12.8   SOFTWARE SHORTCOMINGS

Similar to project management software, if the contractor or owner requires all the team members to use the same scheduling software, linking in all the companies together for viewing and soliciting input requires each of the companies to have access to the software and operators trained in the system. This is not always possible and may be an added expense. The estimator and scheduler should review the special conditions of the contract for software requirements before submitting a bid or proposal.

Many schedulers will start with the computer schedule from a prior project and modify it to meet this project's needs. Although this is much easier than starting from scratch with a blank sheet of paper, it might be too easy. This process might be acceptable for repetitive work, such as highways, utility projects, large tracks of speculative homes, or multistory offices or apartment buildings. But every project is unique and the scheduler must incorporate its nuances into the project plan.

The scheduler must check the computer schedule for obvious errors, similar to checking a computer estimate. One simple method is to visually scan the schedule for the longest and shortest activities and ask the questions, Do these seem right? Can any of the longer ones be broken into multiple shorter activities? Can any short activities be rolled up into one longer activity? Is the arrangement of activities logical? Are any essential but nonconstruction restraints missing, such as material deliveries? Essentially the scheduler should perform a quality control check on the schedule. The critical path should be verified so that activities that are customarily critical, such as foundations, roofing, elevator, electrical switchgear, etc., are on the critical path but others, such as rubber base, toilet accessories, or mirrors, are not shown as critical. Despite all of the technology tools available, the contractor is still responsible and cannot blame the computer. The schedule doesn't make it happen, people do. The computer charts and prints and calculates what has been input by the people involved in the process.

With scheduling software, it is easy to change durations, establish relationships between activities, revise the logic, and create new critical paths, but there are positives and negatives with this. Major changes in the schedule assume that the original schedule was not correct to begin with. When schedules were all drawn by hand, it was very difficult and expensive to issue a new schedule and builders would work hard at finding a way to make the original plan work. Software also makes it almost too easy to update and revise and print a new schedule. Some project owners require a revised schedule on a regular basis, such as monthly. It is important for the baseline schedule to be maintained. The GC's scheduler must understand that the schedule is a contract document and any change may result in a claim for impact or additional time from subcontractors.

The project may be of such a size or the software may be so complex, that a full-time scheduler as discussed in the first example in this chapter is required. It is imperative that the software and scheduler remain tools and not obstacles. The following quote was provided by a real estate developer when researching this book.

> *The reality is scheduling and project management software gets to be so labor intensive that you feel like you are working for the software rather than it working for you.*

The schedule is a tool; it should not take on a life of its own. The scheduler should not be running the office or field side of construction; that is for the PM and superintendent, respectively, whether the scheduler is a third-party consultant or home office specialist. As the next example highlights, not all built environment participants are as equipped as others to manage scheduling software systems.

Example 12.2

A young superintendent had just finished taking a day-long scheduling seminar and was determined to produce his first schedule for this project. He had good intentions, but unfortunately was not adept with the computer. He would spend the entire day in the trailer struggling over the schedule when he was really needed out on the jobsite. The subcontractors worked on what they chose, and their weekly reports were so different from what his printed schedule showed, that he was constantly playing catchup, rather than using the schedule as a forward-thinking planning and control tool. His struggles continued for months, yet he was reluctant to focus on planning and delegate the computer input to his project engineer. The project would eventually suffer significant quality and safety issues and was prone to material theft. The field boss needs to be in the field.

12.9 SUMMARY

Very few detailed construction contract schedules are drawn by hand anymore. There are many popular schedule software programs commercially available today including Microsoft Excel, MS Project, Oracle's Primavera P6, Touchplan, and others. There are advantages and disadvantages to each of the systems and many have similar features, in addition to integration with other project management systems. There is not a 'best-of' judgment made in this chapter, or in this book, regarding the ideal construction scheduling software. The answer is "It depends." With all of the changes in technology available to construction professionals, along with changes in delivery methods, it is difficult to analyze concepts like planning and scheduling and come up with only one solution. The right software solution for any particular contractor is one that can be adapted for future scenarios. Scheduling software has come a long way, as discussed in the next example and it continues to evolve today.

Example 12.3

When this author was an undergraduate student taking the only scheduling course offered on campus through the civil engineering department, schedule technology was just breaking the surface. The process at the time was called Fortran. It involved punched holes in a card of paper which represented a unique activity and included its title, duration, and predecessors. All of the activity cards would be input into a computer the size of a small house. It would take all night, and if there was one card with an error, I would be back the next evening to give it another try, and had to pay extra for it.

This chapter concludes Part III of this book, which focuses on schedule development. These chapters advanced all of the planning efforts into the production of a schedule document that now can be utilized to manage many facets of construction management. Durations and timelines were added to the schedule logic and a critical path was determined by a series of forward and backward pass calculations. Resources such as manpower, equipment, and cashflow are limited in construction and the scheduler must incorporate them into the project plan.

In the next part of the book a variety of construction management control tools and processes are explored, which are direct byproducts of schedule planning and development. The ultimate success of the construction project will rely on these controls. The topics included in Part IV include:

- Schedule control;
- Jobsite controls including cost, quality, safety and others;
- Schedule tools including the submittal schedule;
- Subcontract management including the expediting schedule; and
- Schedule impacts including claims.

12.10 REVIEW QUESTIONS

1. Why should manual scheduling calculation techniques be taught?

2. Why aren't all construction projects scheduled utilizing computer software?

3. What are the advantages of a full-time staff scheduler located in the home office who prepares all of his or her company's schedules?

4. When might a full-time scheduler located at the jobsite be required and what are the advantages of a full-time scheduler located at the jobsite?

5. What are the advantages of a corporate scheduler spreading his or her time between multiple projects, say one day every two weeks on a particular project?

12.11 EXERCISES

1. Other than the software systems discussed in this chapter, what systems have you had the opportunity to work with?

2. Several advantages of scheduling software were noted above. What benefits have you also experienced? Are there potentials for further advancement?

3. Other than the disadvantages of scheduling software discussed in this chapter, what drawbacks have you observed? Did you disagree with any that were noted here?

4. This book advises the PM and superintendent should be intimately involved with schedule development, or maybe they are also the project schedulers. (A) Can all PMs and superintendents perform this work? (B) Prepare an argument why they should not be the project schedulers.

Part IV

Project Controls

Part IV

Project Controls

Chapter **13**

Schedule Control

13.1 INTRODUCTION

The plans developed during the preconstruction period as discussed in Chapter 3 are not useful tools if contractors do not implement the plans, and when situations arise, deviate from them and readjust. The previous group of chapters developed construction schedules and other communication tools to communicate preconstruction plans and schedules to the construction management (CM) team. This last part of the book discusses many management or control tools that contractors utilize to keep the project on course and take advantage of the hard work that went into developing those plans.

It is customary to use the word *control* when referring to cost and schedule controls. Control is loosely defined as to have power over a person or situation. But general contractors (GCs) do not have true power; instead they utilize their tools to manage craftsmen and subcontractors to achieve their desired goal of a successful project. *Manage* is likely more correct than *control* when it comes to construction management concepts, but this book utilizes both terms. The broader term *manage* compared to *control* correlates to several descriptive and beneficial phrases, including:

- Guide, test, verify, check,
- Handle with a degree of skill,
- Treat with care,
- Use means available to accomplish a goal,
- Executive skill and others.

A construction manager at any level must have the personal *ability* and have been given the *authority* to allow other team members to be successful. Both of these are proven measures of an individual's personal leadership skills. A series of control and management tools are utilized by the jobsite supervision team to achieve project success. Schedule control is one of the primary focuses of this book. Project control systems for both the project manager (PM) and superintendent include:

- Safety control,
- Cost control,

- Quality control,
- Document control, and
- Schedule control.

The first four of these and other construction management control systems will be discussed in Chapter 15. A figure is also included with that chapter that reflects how these CM control systems interplay. Schedule control involves monitoring the progress of each activity in the construction schedule and determining the impact of any delayed activities on the overall completion of the project and is the focus of this chapter. This is often the responsibility of the individual who can most "make it happen," the GC's project superintendent. Schedule control is just as important as cost control, because the project team wants to ensure substantial completion is achieved prior to the required contractual completion date. If this does not occur, liquidated damages (LDs) may be owed to the project owner to compensate for construction not being completed on time. Liability for LDs ends when project substantial completion is achieved.

13.2 SCHEDULE CONTROL TOOLS

The general contractor's superintendent should report construction progress at the *weekly owner-architect-contractor (OAC) project coordination meetings* and identify the causes of any delays. He or she should select appropriate mitigation measures and present plans for recovery to the project team, if necessary. Such measures may include expediting material delivery, increasing the size of the workforce, or working extended hours. But if the superintendent cannot communicate his or her plan to the project team, their leadership will be compromised as shown in the next example.

Example 13.1

The schedule is an important communication tool, and its presentation, in whatever format (summary, contract, three-week), needs to be understood by all of the stakeholders. This particular superintendent would spend the entire Monday morning preparing for the foremen's meeting by updating his schedule, but he always seemed to have trouble formatting and printing the schedule. For this particular meeting he printed out and copied for the whole team his 12-page 11 × 17 schedule (this already is not a good communication tool), but he had not proofread it. Half of the pages were blank, others had only vertical restraint lines with no horizontal activity lines or activity descriptions, and the schedule was plagued with reverse logic loops and incorrect dates. The superintendent loses credibility with his or her team if they cannot communicate the importance of achieving the project schedule.

Pull planning is a lean construction skill that involves the individuals who are responsible for accomplishing the work, subcontractors and foremen. Lean construction has aptly labeled those individuals as the "last planners." Many lean planning aspects are discussed in detail in Chapter 5. Short-interval schedules are also developed by the foremen to manage all self-performed work. Specialty schedules include building areas or phases or for individual subcontractors and are also valuable schedule control tools of the jobsite management team.

The *schedule development and control* cycle is shown in Figure 8.1. The first schedule will be developed during the estimate or bid process to help establish anticipated general conditions costs. Schedules are often required by clients with negotiated requests for proposals as well. Even contractors bidding unit-prices for heavy civil projects with owner-provided end dates and LDs will rough out a schedule to determine the project's feasibility. But all of these schedules are placeholders until the contract is awarded.

Once the contractor is notified of intent to award by the project owner, a *detailed contract schedule* should then be collaboratively produced. A construction team cannot perform schedule control to an incorrect plan; they must correct the schedule to make it easy to monitor and control, as will be discussed later. One of the steps necessary to successfully prepare a detailed construction schedule includes an analysis by the project superintendent. The superintendent will input his or her expected work plan and work flow, and will perform manpower and crew size analysis. In addition, after subcontractor and supplier buyout, a collaborative approach to scheduling involves obtaining expected durations and material and equipment delivery dates from the GC's major second-tier team members; their subcontractors. These dates should then be incorporated into the detailed project schedule.

Some superintendents and foremen have a fear of sharing their plan; they may be criticized if they miss a date. That fear may cause a contractor to have one schedule they share with external stakeholders, such as the owner and the designer, and another that they build to. Having two different schedules is akin to two different sets of books in construction accounting, jobsite diaries, or a pay request schedule of values. It is not a good practice and if caught, destroys credibility. A superintendent who keeps his or her plan hidden from other stakeholders is not an effective leader as was exhibited in a previous example.

Schedule control is the *superintendent's responsibility*, with assistance from the project manager (PM) and staff scheduler, if necessary. All members of the built environment community feel that control of the construction schedule is an essential construction management skill. Planning, preparation, and control of the construction schedule continue to be the focus of this book.

13.3 SCHEDULE CONTROL TECHNIQUES

There are many reasons the jobsite management team should pay close attention to schedule control. One of these is if an LDs clause has been written into the prime contract. Liquidated damages may be as low as $100 per day or as high as several hundred thousand dollars per day. Civil projects may have hourly LDs, if not by the minute, for road closures, and the amount of money increases exponentially the longer the violation. Another reason for finishing a project on time is the adage that "time is money." In construction, the jobsite general conditions estimate is largely time dependent. The longer the project lasts, the more general conditions costs are spent, which reduces fees and eventual profit.

Schedule control is more than just the production of reports or discussing the schedule at the foremen or owner-architect-contractor coordination meetings. Schedule control involves careful review of the plan and expected standards, checking for deviations from the plan, and making corrective adjustments. Multiple reports may be generated by a computerized scheduling program, but the computer does not control the schedule, people do. Some GCs have a reputation for achieving schedule success more than others, which can result in more cost-competitive initial bids and increased profits. In addition to the GC's project superintendent reviewing the schedule status at the weekly OAC meeting, the contractor will hold weekly foremen and/or subcontractor coordination meetings, at which schedule, as well as safety and quality and other communication issues, will be discussed.

If the project is behind, and the superintendent must maintain a rigid completion date, he or she cannot (or should not) simply squeeze activities, overlap activities, shorten durations, or increase manpower. It is assumed that the initial plan had some good forward-thinking involved in its preparation before the project started. If these types of changes can simply be made to the original schedule without impacting cost or safety or quality, then the original plan was in error. If the schedule is this flexible, then its use as backup for a change order, back charge, claim, or development of future schedules is negated. But things can change and new ways of building the project may present themselves and the schedule may warrant a formal revision. However, sometimes the project is behind schedule because anticipated productivity was not realized. Overtime or weekend work may be needed to get the project back on schedule.

General contractors are always on the lookout for schedule actions from the project owner, or its agents, which cause major impacts to their plan and costs. Some schedule impact terminology includes:

- Differing site conditions,
- Schedule delay,
- Schedule acceleration,
- Schedule compression,
- Disruption of schedule, and/or
- Suspension of schedule.

If any of these occur, the general contractor must timely notify the project owner of their impact. But the GC's superintendent must also be aware of how his or her company's actions affect subcontractors. Subcontractors today are sophisticated and they also have superintendents who know how to interpret a schedule, and PMs, corporate officers, and attorneys who all know how schedule impacts affect their contract and their original bid. Time extensions on anyone's part affect all contractors' general conditions estimates and potential fee forecasts. They all read books and take classes just as the GC's team does, so it would be naïve for a field manager to believe he or she can pull a fast schedule revision over the subcontractors' eyes and not have them present a claim in return. Subcontract management and schedule delay claims are discussed in Chapters 17 and 18 respectively.

13.4   CONTRACT SCHEDULE: STATUS, UPDATE AND/OR REVISE

Just as floorplans, shear wall schedules, and technical specifications are contract documents, so should the detailed construction schedule be a contract document. The schedule document itself should be referenced and tied to the prime agreement as an exhibit and contract document. The contract schedule is more than simply start and completion dates and construction durations, as is done in many prime contract templates. The contract schedule is a physical (or maybe electronic) document. This same schedule should also be reflected in all subcontract agreements and major purchase orders.

To *status*, *update*, and/or *revise* the detailed contract schedule are maintenance activities involving the jobsite management team and occur throughout the course of construction. Many members of the built environment, including project owners, may see these three terms as the same, but that is not necessarily true. This section will explore the sometimes subtle difference between the three terms and will introduce the development of a recovery schedule that is prepared when there has been a major schedule slip but the end date must still be maintained. Schedule control, by this author, is an active process requiring involvement from the entire team. Other passive terms may be included in a conversation of schedule control, such as monitor, track, observe, report, or measure. And although these terms do play a part, especially in the next subsection on schedule status, they are all passive terms, and not active schedule control. Schedule control as defined here is more than "record" and "monitor," but requires the team to adjust and implement changes if necessary. The Fire Station 83 case study special conditions section 013000.1.5.C.2.a seems to cover all three of these terms in one paragraph titled "Contractor's construction schedule," which reads in part:

> *Review progress since the last meeting and review the three-week look-ahead schedule . . . determine activities which are on-time, ahead, or behind . . . and if behind, how does the contractor propose to expedite the schedule . . . and if revisions to the schedule are necessary, does the contractor still ensure the project will be completed on time?*

Schedule Status

Statusing a schedule is basically a report on current progress or lack of progress. The first step in status requires the monitoring and recording of construction activities: Did the formwork start? Was the reinforcement steel delivered? Is the concrete pour still scheduled for Friday? Has the vibrator been repaired or is there a new one on order? This is essentially a report on current progress. The ultimate goal of the schedule status is to collect and compare data to the original plan and share that report with other project stakeholders. This was reflected as the second step in the schedule control cycle in Figure 8.1.

The date of the schedule status is known as the "data date," which is a snapshot in time. The work has likely already progressed past the data date by the time it is reported. The time or date the schedule is statused is denoted by drawing a heavy vertical line on the schedule that reflects the data date. Activities to the left of the status line were scheduled to have been completed or progressed to the point of the date of the report. The activities to the right of the vertical line are activities that are scheduled to happen or be completed in the future.

The project superintendent should be the individual at the weekly owner-architect-contractor meeting who statuses the contract schedule posted to the trailer conference room wall. An old technique of dropping a plumb bob to status activity progress with respect to today's date is still a useful tool, as shown in Figure 13.1. An interesting anecdote is that the plumb bob used to be an integral tool for the site survey process. But now with lasers and total stations and computers, it is not used for surveying as often. However, it has found its way into the jobsite trailer with a new life. A useful approach for the superintendent is to stand and address project progress including items ahead or behind and any adjustments under consideration to the project plan. The detailed contract schedule for the fire station case study project, which will be reported on weekly by the superintendent and posted to the jobsite trailer meeting room, is included on the book's companion website. Unfortunately, not all superintendents address the schedule on a weekly basis, as shown in the next example.

Example 13.2

This school district had built many elementary and middle school projects successfully but had never built a $100 million new high school from scratch. The project was awarded on a lump-sum basis to an out-of-state GC who had built several schools. As with many projects, there were issues on the owner's, architect's, and contractor's sides that were impacting the schedule. A construction consultant was tasked with observing the weekly OAC meeting and reporting back to the school board about the progress of the high school. The GC's superintendent did not have a schedule posted to the meeting room wall and would not discuss schedule issues at all during the weekly OAC meeting because "everything was all messed up." The contractor blamed the owner for all of the schedule impacts and for not statusing the schedule; contractors were also hiding their own lack of progress.

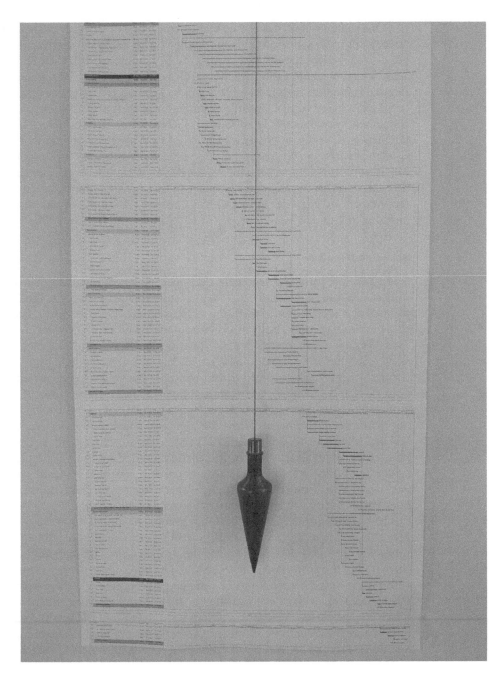

Figure 13.1   Plum bob status example

One simple way to status the detailed construction schedule posted on the jobsite trailer's conference room wall is to highlight with green what activities have started, progressed (relative percentage complete), or completed. It is then easy to see which activities are more or less on schedule. It is best not to use a red highlighter for those that are behind, as then next week it will be difficult to status and update. Notes can be added in hand to the schedule documenting when activities actually started, were completed, materials delivered, and work stoppages. The complete result, whether the project is "right on" (which it rarely is) or is ahead or behind by days or a week is relatively easy to see based on the vertical status line and the green highlights. Maintaining the original schedule as long as possible in this manner keeps the original record, records actual dates (which can be annotated on the schedule) and keeps the heat on: "We haven't given up the ship yet!" A sample of a statused schedule is included in Figure 13.2.

Schedule Updates

Schedules are working documents that need updating as conditions change on the project. The update is one more degree involved than simply comparing actual to planned and preparing or presenting a status report. Updating the schedule for some is the same as revising, but not necessarily so. When project owners or construction executives ask the superintendent to "provide me with an update," they are not necessarily saying print me a brand new schedule, but rather report on how the project is doing, which is really similar to status above. If the field supervisor responds that the original plan is not being accomplished, management will ask in kind, "So what are you going to do about it?" which requires an update of the schedule.

Schedules, like estimates, provide a map of where the project team wants to go and basically how they are going to get there. The objective is not necessarily to complete each activity exactly within its scheduled duration, but to complete the overall project within the scheduled time. Many activities will be completed early, and others late. The PM and superintendent just want to keep from finishing the entire project late. The objective is for the variances to average out.

Just as the project manager and superintendent should not work off an inaccurate cost estimate, they should not use an incorrect construction schedule. The construction team must first understand what the problems are and then endeavor to correct them. Often the answer is going to be that additional information is needed, meetings with project stakeholders are to be arranged, and potentially the schedule document must incorporate additional activities or the scheduler revises the logic. Just as estimates are not perfect, neither are schedules, and perhaps the original schedule was in error. More is known now about the project and corrected schedule logic is often an appropriate step. This may be accomplished through a variety of methods including:

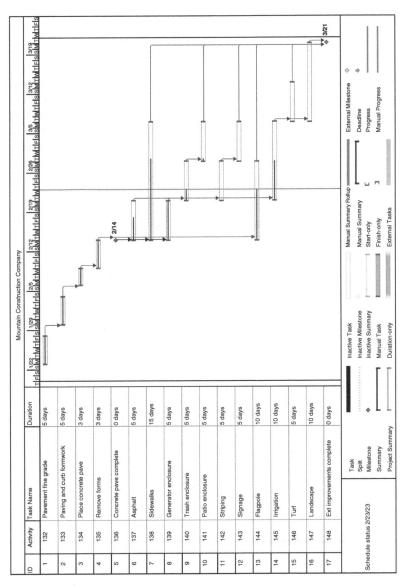

Figure 13.2   Schedule status example

- Paying extra for expedited material deliveries, such as air-freight;
- Increasing manpower;
- Working double shifts;
- Working overtime;
- Changing craftsmen or field supervisors; and/or
- Removing and replacing a subcontractor or supplier.

A schedule update incorporates minor revisions to get the project back on plan. It does not change the end date and not typically the original critical path. For example, contract change orders generally modify some aspect of the scope of work, requiring an adjustment to some part of the schedule. Some activities may be significantly behind and others ahead and a new plan is required, but not necessarily a new completion date. Most of these remedies are not without increased costs and other impacts and risks, including production inefficiencies. The project team must carefully assess the impacts of each method selected. Schedule recovery usually can occur only by accelerating activities on the critical path. In addition to aiding management of the project, updated schedules also can be used to justify additional contract time on change orders and claims. The below example was contributed by a public works owner's representative regarding schedule updates.

This author and industry professional practices the "Line down the middle of the road theory," which relates to both cost and schedule control. Essentially the baseline estimate or schedule is a map; they both reflect what the project team hopes to achieve by project completion. Both the estimate and schedule are composed of many (likely hundreds, if not thousands) unique activities. The contractor's PM and superintendent will not hit every activity perfect such that they perform exactly as scheduled and as budgeted for each event. Some will be finished ahead and others late, some will be finished under and others over budget. It should be the goal of the project team to balance these out. They do not have to follow the line down the middle of the road exactly, it is permissible to veer a little left and right, just stay on the road and get to the destination, or in this case project completion, safely. Minor variances are expected and should be tolerated.

One method to update a schedule is to collapse completed activities and blow up in greater detail the next month's activities. This creates a detailed rolling schedule

Example 13.3

"Look-ahead schedules are typically three weeks out and are presented at the weekly OAC meeting. We also expect the master schedule to be updated monthly so we can address any potential delays that may have occurred and discuss how the schedule losses might be recovered and what activities must be shuffled without adding escalation of the workforce, which can be problematic and expensive."

but still maintains the original contract milestones and logic. An alternative to that approach is to develop several detailed specialty schedules, or fragnets, which supplement but not revise or replace the original contract schedule. An example of a rolling schedule is included in an earlier chapter and a specialty schedule for the roofing subcontractor's scope is included on the companion website.

A lot of time can be put into updating actual start and status and completion dates into the computer, but still some judgment will be required whether the project is ahead, right on, or behind the overall project schedule.

Schedule Revisions

Should the project team ever revise a schedule? If so, when and how? This depends upon contract requirements as well as project progress. If the project is proceeding more or less on schedule, the schedule need not be revised. If the project is significantly behind schedule or there have been many change orders, the schedule should be revised. Current computer scheduling software allows revisions to the schedule to be saved. This provides the capability to compare the current schedule with the original baseline schedule, allowing the scheduler to document schedule impacts that facilitate the PM's ability to negotiate change orders.

The original schedule should be maintained intact if at all possible. If a revised schedule is developed, it must be submitted to all subcontractors to be evaluated for impact. Upon acceptance, the new schedule must be incorporated into every contracting and supplying firm's respective contracts. Similar to incorporating new drawings into contracts, this is a change order opportunity for subcontractors.

Sometimes it is necessary to revise the schedule, either because of a contract requirement or because of a substantial change in scope or progress. It is recommended that some system be used to retain the original commitments of all of the parties. If it appears that construction progress is significantly behind, it may be necessary to develop a new schedule. This may have been caused by a variety of reasons, including:

- Late material deliveries,
- Owner-induced delays,
- Architect-induced delays,
- Weather,
- Additional scope,
- Labor shortages or inefficiencies,
- Equipment availability,
- Inadequate estimate or schedule, or
- Incorrect schedule logic.

Issuance of a completely new revised schedule should be avoided until absolutely necessary because it is a contract document. Just as is the case with a revised foundation plan, a new schedule will result in change orders. Sometimes it is not the contractor's choice if a new schedule revision is published. Prime contract terms often describe schedule processes, including format and revisions. Some project owners require a new schedule to be issued monthly, often to accompany the pay request. Some change orders are so significant that they also require a revised schedule be produced. Revising and reissuing schedules monthly is not without a cost.

A subcontractor could make the argument that all changes made to a contract schedule should be clouded the same as changes that are made to a design drawing. One option would be for the scheduler to provide a list of all of the changes made, including logic, durations, and dates, to make it easier for subcontractors to understand and incorporate the new schedule into their work plan – scheduling software can accomplish this. When a schedule is revised and reprinted and distributed, it is imperative that a new revision date be posted on the schedule, similar to a revised drawing, such that all stakeholders are working off the same, most current document.

Recovery Schedules

There are ways to revise a schedule for a project with a hard fixed end date, such as a school opening by September. Some of them include: Reduce durations, increase crew size and quantity, stack construction crews, add overtime, compress activities into a shorter duration, add new logic, draft a new plan, and change predecessors and/or relations. But as stated previously, most of these have cost ramifications and other impacts and risks, including production inefficiencies.

In the case of a necessary schedule revision due to delays caused by the contractor, such that the end date must still be achieved, is it as simple as finding new logic? If so, then it can be argued that the original logic was either not correct or was not the most efficient plan. A recovery schedule typically involves more than revised logic. Often the solution to prepare a new recovery schedule with a fixed end is to "crash" the schedule, which implies accelerating or compressing all activities. Builders typically avoid using the word "crash," as it has negative connotations and would likely trigger subcontractor complaints and notification of claims due to loss of productivity.

If a project is delayed, and the jobsite team must make up the time, shortening or crashing or compressing the entire schedule, or even just one activity, is not a simple task and may come with risks. This may create other critical paths, which have cost implications. The scheduler should perform a cost-benefit analysis when revising the schedule. The project team must carefully assess the impacts of each method selected. Schedule recovery usually can occur only by accelerating or compressing activities on the critical path. A recovery schedule must also be submitted to all subcontractors to be evaluated for impact on costs.

But if the scheduler shortens one critical path another new critical path will likely be created or kicked in. To make up for lost time or accelerate the schedule, many scheduling books recommend the scheduler simply reduce durations. This method is not the correct approach as it implies the activity durations were inflated to begin with or there is no limit to the density the crew can work without a loss of productivity, as is discussed in Chapter 10. A superintendent should be consulted before any work activities or paths are reduced in duration.

A better use of a recovery schedule is to choose specific tasks or one path to accelerate; this is a more productive method than crashing or accelerating the whole project. Not all activities are critical and blindly requiring the entire field team to work overtime is an expensive effort as shown in Example 10.2. The best solution therefore is to work around the problem and come up with a better plan.

13.5 THREE-WEEK LOOK-AHEAD SCHEDULES

While contract schedules, such as those discussed earlier and in other chapters, are developed for overall project control, short-interval schedules generally are used by foremen and subcontractors to manage the day-to-day activities on the project. These short-interval schedules are developed each week throughout the duration of the project. Three-week schedules typically provide sufficient information for managing the project from a field supervisor's perspective. Subcontractors should be required to prepare and submit to the project team their three-week schedules at the Monday-morning foreman's coordination meeting. The project superintendent then collects all of the subcontractor schedules and summarizes and presents his or her three-week schedule at the weekly OAC meeting.

One of the first short-interval schedules a general contractor will create is to encapsulate all of the mobilization activities required before construction begins. The initial three-week schedule must be prepared during project start-up to schedule the GC's and subcontractors' workforces to minimize interference and smoothly plan the flow of work. Development of the initial three-week schedule for the fire station case study project is included as a figure in Chapter 1.

The short-interval schedule may be referred to as two, three, or four-week schedules, look-ahead schedules, or foremen schedules. More important than the format or title of the short-interval schedule is its author. The next example is a unique use of a short-interval schedule communicating the GC's plan at the weekly OAC meetings. The preparation of short-interval schedules is the responsibility of the last planners – the foremen, subcontractors, and superintendents who will accomplish the construction work. It is imperative that they have buy-in with the schedule. The schedules should not be delegated to the project engineer, but the project engineer may assist the builders with computer scheduling tools. Some projects are also choosing to utilize team-developed pull schedules as their short-interval scheduling tool.

The detailed three-week schedule is a powerful tool of the last planners, including superintendents and foremen. One unfortunate method utilized by some PMs to respond to the requirement to submit a weekly look-ahead schedule to the project owner is to simply take a three-week window from the detailed computer schedule without adding any additional updates or detail, and hand that out at the weekly

Example 13.4

A local general contractor experienced at cast-in-place concrete work and parking garages negotiated a 3,000-stall seven-story garage for this casino project. Even though the project was negotiated, the project owner still required a $100,000-per-day liquidated damages clause in the contract if the garage was finished late; they needed their parking stalls to support a grand opening! The garage had two distinct halves with ramps up and down through the center. The contractor utilized two tower cranes and two concrete placing booms. The owner was not experienced with construction projects and did not really understand the schedule. The superintendent was a good communicator and built a physical scaled model of the garage. He would explain with the model in the weekly OAC meeting what half and what floor was being poured that week and which was being shored and reshored. These activities all corresponded to the contract schedule. This was the best three-week schedule the GC could have used and was accepted by the owner. The project finished on time and was a big success for all of the parties.

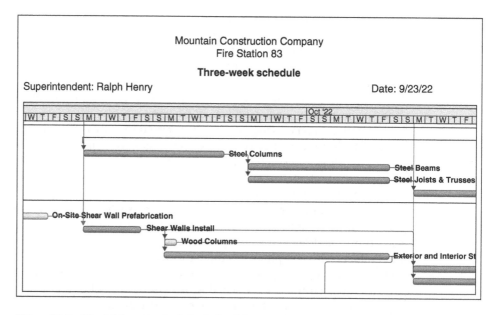

Figure 13.3 Short-interval schedule derived from contract schedule

OAC meeting. Figure 13.3 is an example of just such a schedule from the case study project. This is not a construction tool. If all was exactly on schedule per the original plan (which would be unusual), then these activities are a good start, but considerable additional detail is necessary, especially input from subcontractors and foremen and suppliers. This example was prepared utilizing Microsoft Project scheduling software.

Another similar example of a six-week look-ahead schedule printed straight from the entire project schedule is included as Figure 13.4. On that two-year project the scheduler simply requests a six-week sort and does not even change the calendar from years and months to weeks and days or adds any additional detailed activities. This is not a good tool to communicate with the last planners to facilitate construction of the project. This example was prepared utilizing Primavera P6 scheduling software.

Short-interval schedules are the link between a contract schedule and construction operations. These communication tools bridge the gap between project owner and jobsite management discussions versus those of craft supervision including foremen and subcontractors. Not all of the detail can be put into a contract schedule, it would be too cumbersome, too difficult to maintain, and would not be an effective communication tool with stakeholders above the foreman level. Some owners may request copies of the contractor's three-week schedules and then hold it over their head if they missed a day here or there. This should not be the case and may cause some superintendents to hold information back or only provide schedules similar to Figures 13.3 and 13.4. Owners should be looking at the short-interval schedule as a tool to also help with their planning; for example: What approvals or decisions are needed from the owner in the short-term? What inspections should the owner schedule? How does the contractor's operation affect their ongoing operations?

Short-term scheduling is one of the primary means of communication for a superintendent and foremen and subcontractors. Another term for this schedule control tool is production schedule or short-interval production schedules, but construction is different than production industries, such as car manufacturing. Speculative home construction or prefabricated modular construction may have more similarities

To-Do Lists

To-do lists are also great short-term scheduling tools. To-do lists should not be overly full or long as each jobsite manager must have room and flexibility to deal with today's unknowns or fire drills. An alternate to a to-do list is a simple calendar that can be generated by scheduling software as an alternate view or print-out compared to bar charts or precedence diagrams. Examples of both a to-do list and a calendar schedule from the case study are included on the companion website. The next example reflects a union of a to-do list and a short-interval schedule.

		Actual Level of Effort	Actual Work		Remai...	6 Weeker
L2PT-1150	Cure Time for Stressing	3	17-Jul-20	21-Jul-20		Cure Time for Stressing
L2PT-1180	Stress PT Slab Tendons	1	21-Jul-20	21-Jul-20		Stress PT Slab Tendons
Level 2 Walls and Columns		16	13-Jul-20	03-Aug-20		Level 2 Walls and Columns
L2 Walls Stair 2		11	17-Jul-20	31-Jul-20		L2 Walls Stair 2
L2WL-1000	Confirm Controls and Layout	1	17-Jul-20	17-Jul-20		Confirm Controls and Layout
L2WL-1010	Set Core Formwork @ L2	1	20-Jul-20	20-Jul-20		Set Core Formwork @ L2
L2WL-1030	Install Rebar	2	21-Jul-20	22-Jul-20		Install Rebar
L2WL-1060	Install Embeds	1	23-Jul-20	23-Jul-20		Install Embeds
L2WL-1070	Install Sleeves	1	24-Jul-20	24-Jul-20		Install Sleeves
L2WL-1090	Rebar Inspection	1	27-Jul-20	27-Jul-20		Rebar Inspection
L2WL-1120	Close Forms	1	28-Jul-20	28-Jul-20		Close Forms
L2WL-1150	Place Concrete at Core Wall	1	29-Jul-20	29-Jul-20		Place Concrete at Core Wall
L2WL-1190	Strip One-Side	1	30-Jul-20	30-Jul-20		Strip One-Side
L2WL-1220	Jump Core Formwork	1	31-Jul-20	31-Jul-20		Jump Core Formwork
L2 Walls Elev 1&2		11	20-Jul-20	03-Aug-20		L2 Walls Elev 1&2
L2WL-1020	Confirm Controls and Layout	1	20-Jul-20	20-Jul-20		Confirm Controls and Layout
L2WL-1040	Set Core Formwork @ L2	1	21-Jul-20	21-Jul-20		Set Core Formwork @ L2
L2WL-1050	Install Rebar	2	22-Jul-20	23-Jul-20		Install Rebar
L2WL-1080	Install Embeds	1	24-Jul-20	24-Jul-20		Install Embeds
L2WL-1100	Install Sleeves	1	27-Jul-20	27-Jul-20		Install Sleeves
L2WL-1130	Rebar Inspection	1	28-Jul-20	28-Jul-20		Rebar Inspection
L2WL-1160	Close Forms	1	29-Jul-20	29-Jul-20		Close Forms
L2WL-1200	Place Concrete at Core Wall	1	30-Jul-20	30-Jul-20		Place Concrete at Core Wall
L2WL-1230	Strip One-Side	1	31-Jul-20	31-Jul-20		Strip One-Side
L2WL-1260	Jump Core Formwork	1	03-Aug-20	03-Aug-20		Jump Core Formwork
L2 Walls Stair 1		11	13-Jul-20	27-Jul-20		L2 Walls Stair 1
L2WL-1110	Confirm Controls and Layout	1	13-Jul-20	13-Jul-20		Confirm Controls and Layout
L2WL-1140	Set Core Formwork @ L2	1	14-Jul-20	14-Jul-20		Set Core Formwork @ L2
L2WL-1180	Install Rebar	2	15-Jul-20	16-Jul-20		Install Rebar
L2WL-1250	Install Embeds	1	17-Jul-20	17-Jul-20		Install Embeds
L2WL-1270	Install Sleeves	1	20-Jul-20	20-Jul-20		Install Sleeves

Legend: Remaining Level of Effort — Actual Level of Effort — Actual Work — Remai... | TASK filter: 6 Week Lookahead

Figure 13.4 Six-week short-interval schedule example

Example 13.5

When this author was still in college and working summers for his father as a carpenter foreman, he had the opportunity to prepare a very simple short-interval schedule, without knowing he was doing so. The builder was going to be away from the current house project for two days but gave his son a list of activities that he was to work on. This included: damp proofing the back side of the recently poured concrete retaining wall, removing snap ties from the wall, installing the perforated drain system behind the wall, installing the pressure-treated plate on top of the wall – which included drilling for all of the cast-in-place anchor bolts, and cleaning a large pile of form lumber from nails and concrete and form oil so that the material could be used for framing, scheduled to start the following Monday. The only caveat the builder gave the young carpenter was to (A) keep busy, and (B) be prepared for the backhoe that is coming at 7:00 a.m. on the coming Thursday to backfill the wall. The young carpenter prioritized the work and roughed out the following to-do list/schedule for himself. The next term at school he would use this as a homework assignment in his scheduling class and his professor asked his permission to use it as an example in his textbook.

- Tuesday: Remove snap ties, apply the first coat of damp proof and let dry, begin cleaning forms.
- Wednesday: Apply the second coat of damp proof, install the footing drain, continue cleaning forms if not complete, begin pressure-treated plate layout, drilling, and installation.

13.6 SUMMARY

Just as a broken hammer or dull saw are ineffective construction tools, so are inaccurate estimates and schedules. These documents need to be fluid to be effective construction management tools. The cost and schedule control processes are circular and the starting point depends upon an individual's experience and the longevity of the construction company. The steps in this control process, as depicted in Figure 8.1, include: (1) Schedule, (2) Correct, (3) Record, (4) Modify, (5) As-build, and back to (1) Schedule. Changes to the estimate and schedule begin shortly after the project is bid, when errors are corrected and buyout commences.

Schedule control is a topic the construction industry feels belongs in any construction management book. More than any other individual, the GC's project superintendent is the one team member who is responsible for achieving schedule goals. Statusing the schedule entails reporting progress and any revised planning necessary to accommodate deviations. The project superintendent often stands and addresses the detailed contract schedule in the weekly OAC meeting and reports on schedule status. Using the schedule to monitor progress is the superintendent's responsibility. Stuff happens on jobsites and it is up to the jobsite management team, including the superintendent, PM, and/or scheduler to be able to adjust to changed conditions. A project engineer can assist by contemporaneously annotating the contract schedule that will contribute input to the as-built schedule during the close-out process.

Updating or revising the schedule can be an expensive task and does not need to be a monthly occurrence; if the project remains on track, leave it be. An update of a schedule may be necessary to incorporate minor changes and work-around plans. In addition to aiding management of the project, updated schedules also can be used to justify additional contract time on change orders and claims. Revising the schedule can be an expensive task and does not need to be a monthly occurrence if the project remains on track.

A schedule revision is necessary when new work has been issued and a major change to the contract schedule must be incorporated. The schedule is a contract document, so any revision requires incorporation into the prime contract agreement and all subcontracts. A recovery schedule is appropriate if the project has a hard, fixed completion date. Crashing or accelerating all activities is not necessary – each path must be carefully evaluated.

The project superintendent also presents his or her short-interval schedule at the owner's meeting. This detailed schedule is a compilation of all of the subcontractor three-week schedules that were exchanged and discussed at the weekly Monday

Example 13.6

The top of this power plant containment dome was 300 feet above the false floor which served as the ceiling for the reactor. A large gantry crane ran on a circular track inside of the dome designed to eventually service the power generation equipment. There were five circular pipes for containment coolant spaced at even intervals high up inside of the dome. The hangers on these pipes required redesign and rebuilding after code changes. The contractor's means and methods included a very tall scaffold on top of the gantry crane that would be moved every five days to access a few additional hangers on either side of the dome. This work was on the critical path and lasted for nearly three months. The mechanical scheduler built a mockup of the crane and pipe loops and hangers for the project's construction manager – a licensed nuclear engineer who oversaw this project's 5,000-person workforce. The scheduler would update the mockup in the CM's office once weekly with work accomplished, including rotating the crane and scaffold to its next position. The CM didn't know many of his 5,000 employees personally, but he knew this scheduler (one of 40 on the project) on a first-name basis.

morning foremen's meeting. There are numerous other specialty scheduling tools available to jobsite management teams, including area schedules and expediting logs. These schedules are also more detailed than the contract schedule. As reflected in the final example for this chapter, schedule control and reporting can take on a variety of useful forms. Several schedule control tools, such as a submittal schedule, are presented in the next chapter. Other construction management control systems will be discussed in Chapter 15, including safety, quality, cost, equipment management, environmental, traffic, and document controls.

13.7 REVIEW QUESTIONS

1. Why should the project superintendent, and not the project manager or project engineer, status the schedule at the weekly OAC meeting?

2. What are liquidated damages? When does a contractor's risk exposure to them end?

3. What is the difference between statusing a schedule and revising a schedule?

4. What is the difference between a three-week look-ahead schedule and a pull planning schedule?

5. Why should contractors avoid revising and reissuing the schedule?

6. When is it appropriate to revise the schedule?

7. How is an as-built schedule developed, and when should it be developed?

13.8 EXERCISES

1. Why did the case study superintendent include non-subcontractor-specific activities on the specialty schedule included on the companion website?

2. Prepare a three-week schedule for the period beginning in early November for the case study project. Include at least twice the quantity of activities already shown on the detailed schedule. Feel free to make whatever assumptions as are necessary.

3. Part A: Prepare a three-week schedule for your coursework starting today. Incorporate deliverables and examinations as milestones, and schedule your classes and study time and personal time. Part B: Status your schedule one week from today. What changes are necessary to finish the remaining two weeks per your original plans? Part C: After week three prepare an as-built schedule. How did you do? Are there lessons learned here that you can apply to the balance of your studies?

4. Start with the fire station detailed case study schedule and incorporate the updates listed here. Part A: Without changing schedule logic or durations, what is the new projected completion date? Part B: Often a GC will get the drywall and painting subcontractors to work on Saturdays to make up for lost time. This may take

some schedule analysis effort from you, but how much time could be gained? What are some of the potential impacts or costs of this proposed revision?

- MEP inspections were completed 12/21,
- Interior doors arrived on-site 1/16,
- Two interior storefront window frames were damaged during installation and will be replaced by 1/20, and
- Casework will arrive two days before scheduled.

5. Start with the network you prepared from Exercise 9 in Chapter 8 and incorporate the updates listed here. Part A: Without changing schedule logic or durations, what is the new projected completion date? Part B: Assume this is from the fire station case study project and propose changes in logic and/or durations and maintain the original schedule completion date. What are some of the potential impacts or potential costs of your proposed revision? Make whatever assumptions as necessary.

- Activity 12 took two days to complete,
- Activities 18 and 20 were able to occur concurrently,
- The project is shut down for two days between activities 24 and 25 due to frozen ground, and
- The fire hydrants will be delivered on June 29.

6. What construction craft uses a level, transit, total station, or laser but no longer uses a plum bob? What union is that craft often associated with?

7. If a general contractor's contract includes liquidated damages, what happens to subcontract agreements?

8. Which subcontractors may warrant general contractor preparation of specialty schedules, and why them?

9. Provide three examples (people) of 'last planners.'

10. Provide three additional activities that could be added to the roofing specialty schedule included on the companion website. Include one for preconstruction, one for construction, and one for post-construction. Make whatever assumptions are necessary.

11. This chapter utilized the three-week schedule format as an example of a look-ahead scheduling tool. Other contractors use two- or four-week durations with similar results. (A) What has been your experience with look-ahead schedule durations, and (B) Which duration do you feel is best and why?

12. Assume all other fire station activities are as-scheduled and utilize Figure 13.2 to determine whether the project is 'right on, ahead, or behind' schedule.

13. Draw a short-interval schedule for Example 13.4. Make whatever assumptions are necessary, including additional activities. Incorporate the backfill and rough framing start-date constraints.

Scheduling Tools

14.1 INTRODUCTION

There are many other schedules that contractors use as tools in construction other than time-scaled bar charts or networks, such as have been the focus of this book. The term *schedule* has many different connotations in construction management (CM). The first group of schedules listed here are design and contract schedules and are out of the contractor's control. The second group are part of the contractor's scheduling tools and are under their control of means and methods of construction. These specialty schedules, and others are the focus of this chapter. For construction purposes, some of these scheduling tools may also be referred to as "logs."

Design and contract schedules:

- Prevailing wage schedule,
- Footing, column, and shear wall schedules,
- Door and door hardware and window schedules,
- Room finish schedule,
- Mechanical equipment and diffuser schedules,
- Lighting and electrical panel schedules,
- Landscape schedules including plant and irrigation schedules, and others.

Construction scheduling tools:

- Submittal and expediting schedules or logs,
- Schedule of values (SOV) for bid submittal or pay requests,
- Concrete pour schedule,
- Earthwork export and import logs,
- Equipment schedule or log,
- Start-up, balancing, and close-out logs or schedules, and others.

Submittal schedules, delivery schedules, and expediting schedules are not necessarily part of the detailed construction schedule, but they back up and support and connect with the detailed schedule. These specialty schedules help resolve when materials are required to be on the jobsite. Poor submittal management may impact the construction schedule and may potentially become critical path activities. Submittal management begins with buyout, submittal preparation, submittal processing, submittal approval, material fabrication, and delivery. All of these critical steps precede construction. A typical rule-of-thumb on the jobsite is if a submittal is turned in late by supplier, late returned from the designer, or, worse yet, rejected, then fabrication will be late, delivery will be delayed, and construction impacted. There are a variety of scheduling tools utilized by both the project manager (PM) and superintendent that are the focus of this chapter. Scheduling reports are also not schedules per se, but they are scheduling tools especially connecting field activities with home office concerns. Reports are included in this chapter as well.

14.2 SUBMITTAL SCHEDULE

The general contractor's (GC's) project manager's and superintendent's primary responsibilities are to complete the project on time, safely, within budget, and to specified quality requirements. The submittal process is a key part of the overall schedule management program for the project. Submittal concepts and processes are the same for subcontractors who supply their own material as they are for the GC.

A submittal is a document or product turned in by the construction team to verify that what they plan to purchase, fabricate, deliver, and ultimately install is in fact what the design team intended by their drawings and specifications. It serves as one last check and validity of design and therefore one of the first quality control (QC) steps. Submittal requirements for a project are contained in the specifications of the contract. Project managers should look upon submittals as an early step in schedule control, and, therefore, as a tool to be used to successfully complete a project. The procedures for processing submittals generally are in Construction Specification Institute (CSI) divisions 00 and 01 of the contract specifications. These special conditions of the contract typically contain specific instructions regarding submittal preparation, copies to be provided, and the review process. The Fire Station 83 case study specification section 013200.2.1.A includes the following requirement for Mountain Construction Company (MCC) to prepare a submittal schedule: "submit a schedule of submittals in chronological order by the dates to be submitted . . ."

Major materials are those that require submittal approvals and have a significant lead time or delivery time. Examples of these include structural steel, concrete reinforcement steel, specialized mechanical or electrical equipment, and many interior finish items. These and other materials typically require designer approval before release of an order for fabrication. The normal procedure is to first select the supplier; issue a purchase order (PO); receive submittals; obtain approval of the submittals from the design team; receive the materials onsite; and store the materials until ready for installation. This is all reflected in the flow chart in Figure 14.1. A member of the

Submittal flowchart

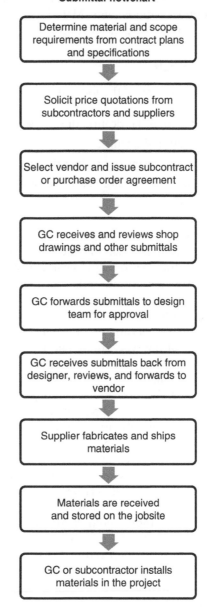

Figure 14.1 **Submittal flowchart**

design team verifies that the submittals, prepared in a variety of formats, correctly interpret contract requirements. Material dimensions based upon field conditions are the responsibility of the general contractor.

Types of Submittals

A submittal can be any of the following:

- Shop drawings, fabrication drawings, or installation drawings,
- Coordination drawings,
- Material cut sheets of product data including safety data sheets,
- Material samples, color charts, color boards, and
- Mockups.

Shop drawings are drawings or diagrams prepared by suppliers to illustrate their products in more detail than included in the contract drawings. Product data sheets are used to illustrate performance characteristics of materials described in shop drawings or submitted as verification that the materials meet contract specifications. Other submittals are used to demonstrate that materials selected for the project conform to contract requirements; these may be performance requirements, descriptive requirements, such as color of carpeting or wall covering, or proprietary requirements.

Regardless of the type, submittals are one of the final design steps and also one of the first schedule control steps. If the supplier is late turning in a submittal, the schedule may be impacted. If the designer is late processing a submittal, the schedule may be impacted. And if a submittal is rejected by either the GC or the designer, the schedule is very likely to be impacted. This is why it is imperative that the contractor's scheduler incorporate submittal relationships into their schedule logic.

Shop drawings are likely the type of submittal that takes the longest to prepare. Structural steel is a material common to many commercial and civil construction projects. Structural steel typically is fabricated by a supplier and erected by a subcontractor. First, the fabrication shop drawing submittal is approved by the GC and the project owner's structural engineer. Then the supplier cuts the steel elements to the correct dimensions and drills the holes for bolted connections. The steel erection subcontractor submits additional shop drawings showing the erection sequence. Once these have been approved and material has been delivered, the subcontractor can begin erecting the steel.

Submittal Processing

Submittals also allow the GC to identify some of the hidden errors and exceptions the subcontractors and suppliers have taken in their bids. Although subcontractors will say they bid it "per plans and specifications," and although the subcontract agreement or purchase order clearly reinforces this, the GC's project manager does not want to be surprised three months later when the wood doors that the supplier has delivered are oak veneer and not solid oak. Although the PM may be contractually correct, this does not help get the building turned over, the owner moved in, and retention released.

The submittal process is one of the early checks of the validity of the construction schedule. If the toilet accessory cut sheets are submitted late by the supplier, a good case could be made that the delivery will also be late. The timing of submittals, as well as deliveries and construction installation, should be noted on all of the POs and subcontract agreements.

Submittal planning involves the development of a schedule of submittals that is given to the architect for review and also an expediting log to manage the submittal process with subcontractors and suppliers. Article 3.16 of the ConsensusDocs 500 contract agreement includes six paragraphs describing the requirement for the general contractor's management of submittals. This is typical of many owner and GC or CM prime agreements. Shortly after receiving award of the contract, the GC's PM should review each specification section for submittal requirements. The GC's project engineer (PE) often assists with this process. Language should be included in each subcontract and PO regarding quantity and timing of submittals. The construction schedule is then reviewed with the superintendent and submittals are scheduled. A preliminary submittal schedule with all potential submittals is prepared and often sent to the design team for approval as described earlier.

After a submittal is received, it should be reviewed by the GC's field management team with respect to the requirements in the specifications and the drawings. The PM does not want to be in a position of explaining why the floor covering contractor submitted vinyl tile when slate was clearly required. Vendors have been known to attempt to substitute a cheaper, nonspecified product. The shop drawings should be reviewed for requested dimensions. The submittal should be stamped received, reviewed, approved, or rejected. Each submittal is assigned a number and entered on a submittal schedule or submittal log. This log is updated and discussed at the weekly owner-architect-contractor (OAC) coordination meeting as well as at subcontractor coordination meetings. This is different from the GC's expediting log, which is introduced in Chapter 17. Expediting issues and problems with subcontractors should be kept out of meetings with the owner. Discussion at the OAC meeting should be limited to submittal materials and documents that the designers have received or will receive shortly. An abbreviated submittal schedule for the case study is illustrated in Table 14.1.

Submittals may be processed and approved by the architect, the owner, the city, consultants, engineers, or a combination of the above. They may be reviewed by several people, either concurrently or sequentially. The GC's field management team should set up the submittal schedule to be able to document the entire review process. After receipt of the processed submittal from the reviewing party, the GC needs to review it again for disposition and log it accordingly. If changes have been made to the products by the designer, the PM or PE needs to make note for potential change orders. The submittal should be sent back to the originating party who will

Table 14.1 Submittal schedule

Mountain Construction Company

Submittal schedule

Project No.: 9922   Project Name: Fire Station 83   Project Manager: Charles Kent

No.	Originator	Sub's No.	Specification Section	Description	Date Submitted	Date Return Requested	Date Returned	Disposition*
1	Earth Enterprises	1	Civil drawings	Import sample	6/1/22	6/8/22	6/6/22	A
2	Earth Enterprises	2	Civil drawings	Pea gravel pipe bedding	5/28/22	6/8/22	6/6/22	A
3	Northwest Construction	NA	Civil drawings	Silt fence data sheet	5/31/22	6/8/22	6/3/22	A
4	Northwest Construction	NA	Civil drawings	Perforated pipe sample	6/2/22	6/15/22	6/10/22	AN
5	Steel Fabrication	1	032000	Footing rebar shop drawings	6/7/22	6/15/22	6/21/22	AN
6	Steel Fabrication	2	Structural drawings	Steel embed shop drawings	6/7/22	6/15/22		
7	Rolling Door Co.	1	083613	Fabrication drawings				

* A — approved, AN — approved as noted, RR — revise and resubmit, and R — rejected

then take action. Other subcontractors and suppliers may also need to review the final submittals.

To create the submittal log or submittal schedule, the project manager or scheduler should review the front-end of the specifications. Typically, CSI division 016000 includes a preliminary list of submittals required for the project. Each individual technical specification section then also requires review for required submittals. The GC's field management team should make as long a list of potential submittals as possible. They should exhaust the documents, drawings, and specifications, and review for other typical materials that potentially should be submitted. A missed submittal is similar to a missed schedule restraint. The contractor's PE is likely involved with assembling all of this information to be placed in the initial submittal schedule. One method is to query other stakeholders such as subcontractors, suppliers, the GC's superintendent, and the design and owner teams. After assembly, the PE will send this list, often before schedule dates have been added, to the architect and owner and request their approval, and ask, "Did we get them all?"

Missed or overlooked schedule items often become critical, both for submittals and construction logic. A missed restraint has a more significant impact than if the carpentry crew is off by one day hanging doors. Lack of submittals can become a bone of contention for the project owner with respect to quality control, as shown in the next example.

Example 14.1

This is a small office tenant improvement project with a contract value of approximately $500,000. The owner is very experienced and routinely has one to two projects under construction. The architect is a repeat firm. The owner usually likes to negotiate projects but has recently had two bad experiences with general contractors and decides to bid this project out in hopes of finding a new GC they can develop a relationship with. The general contractor who is selected is also looking for just this sort of arrangement. They are a very small firm. Even this project is at their limit. They operate very informally with little documentation. The project is on schedule and all parties are operating in a partnered fashion. Samples of the carpet are not submitted, as the specifications did not require this process. The carpet arrives two weeks before it is scheduled for installation. It is left rolled up and wrapped. The superintendent does not inspect it. The carpet was installed by the subcontractor over a weekend. The GC's superintendent was not present during this work. On Monday morning the owner arrives and is surprised to see that the new carpet is a different color and a different cut than the carpet it adjoins with and was intended to match. It turns out that the carpet is "as specified" but the architect erred with their specifications. Did anyone "win" in this scenario?

14.3 PROJECT MANAGEMENT SCHEDULING TOOLS

Pay Request

Development of the monthly pay request is an important project management responsibility. An argument can be made that receiving timely payments may be *the most important* PM responsibility. This is not a PM book, but the pay request is integral with the construction schedule, just as it is with the cost estimate.

Estimated costs can be loaded onto the construction schedule as a resource, but then the schedule loses some of its effectiveness as a construction tool. The schedule should not, by itself, serve as a pay request tool. An updated schedule, especially a summary schedule, may accompany a monthly pay request as backup.

A major factor affecting the monthly pay request process is the project's estimated completion status. The detailed contract schedule can help with this. It is difficult to determine the percent complete for the whole job; one cannot simply add up the percentages of completion of the schedule's parts and pieces or systems. The schedule of values that forms the basis for the monthly pay request does this to some regard. A pay request SOV for the case study project is included on the book's companion website. Progress for activities, systems, or areas are typically measured on percent complete, which is calculated according to any of the following methods. Each line item in a detailed pay request SOV has its own percent complete, utilizing one of these calculations.

$$\text{Percent complete} = \text{Hours spent} / \text{Total hours},$$
$$\text{Percent complete} = \text{Cost (\$) spent} / \text{Total estimate},$$
$$\text{Percent complete} = \text{Quantity complete} / \text{Total quantity, and/or}$$
$$\text{Percent complete} = \text{Time elapsed} / \text{Total duration}$$

Monthly Fee Forecast

Just about every general contractor project manager and every subcontractor project manager prepares a monthly cost and fee forecast for their home office. The home office wants to know how costs are looking, in addition to other crucial control areas, as discussed later with reports. The monthly forecast is essentially a new estimate for the entire project each month. The forecast is prepared by adding to-go costs to actual costs and comparing that to the original estimate as reflected in the next formula.

$$\text{Actual cost to date} + \text{Forecasted cost to go} = \text{Forecasted total cost}$$
$$\text{Original estimate} - \text{Forecasted total cost} = +/- \text{Cost variance}$$
$$\text{Cost variance} + \text{Original estimated fee} = \text{Forecasted new fee}$$

The project manager prepares this calculation for each line item in the original estimate. The most difficult element to determine is the forecasted cost to complete.

The PM must evaluate many variables including cost trends to date, learning curves, uncompleted quantities of work, and the plan for to-go work in order to develop an accurate to-go cost forecast for each item in the estimate. The superintendent should assist the PM with monthly cost and fee forecasting. Alternate methods to calculate to-go hours and costs are discussed with earned value in Chapter 16. The superintendent is the best-qualified team member to factor past, present, and future labor productivity into estimated to-go costs and durations.

Float Management

Who owns the float? Read the contract. But even if stipulated in the contract, there remains a debate as to whether the contractor, which implies the GC, or the project owner owns the float. There are many float ramifications involved with construction scheduling, including:

- How does float affect subcontractors and suppliers? Are they also the contractor and if the prime agreement is referenced in subcontract agreements, does the GC then pass the float on down to its subcontractors? General contractors like to pass potential liquidated damages down to subcontractors, so it seems fair they would share the float ownership, but this is unlikely.
- If the contractor plans a shorter duration than the owner's contracted completion date, such as 11 versus 12 months, who owns the float? Many owners think they do, but so does the GC.
- Contractors can create new float by beating the schedule. If the contractor is ahead of its plan, does the owner or design team then have more time for decision making?
- Designers and the city also feel they have a right to use the float with respect to decisions, permits, and inspections.

The solution for many project managers is to not show any float. This is accomplished in a variety of fashions including:

- Introduction of new smaller activities, generally noncritical activities, that consume the float.
- Split longer activities into shorter, often loosely defined activities such as (A) start floor covering, (B) continue floor covering, and (C) finish floor covering.
- Introduce additional front- or back-end activities that may have been considered "assumed" within a longer activity. These would include: layout, cleanup, test, inspection, and others.
- Add long-lead material deliveries, even if they are not critical, or delivery of shelf materials not requiring fabrication or submittals.

As-built Schedule

One type of specialty schedule that is not always developed is an as-built schedule. This schedule is created throughout the course of the construction project by recording actual delivery dates, actual start and completion dates, milestones realized, and potentially subcontractor manpower. The as-built schedule is an excellent element to include with a post-project lessons-learned report. Unfortunately, the as-built schedule may also be utilized in claim preparation or defense – so it is important that only verifiable data be recorded.

Schedules should be maintained in a manner similar to the way record drawings and as-built estimates are maintained. They should be marked up and commented upon to reflect when materials were delivered, activities started, activities finished, rain days, crew sizes, and whatever else the project manager and superintendent feel is appropriate. Heavy civil unit price projects also must record actual material quantities installed. The as-built schedule, similar to the daily job diary, often becomes a legal tool in the case of a dispute. The recording of actual durations can also be used to more accurately develop future schedules and should be shared with the home office staff scheduler.

Photographs are good tools to record as-built schedule status, especially if they are date-stamped. Photographs are also good construction communication tools to accompany requests for information, change orders, meeting notes, and others.

14.4 SITE SUPERVISION SCHEDULING TOOLS

An example of a *tabular schedule* is shown in Figure 7.1. Some contractors, and especially superintendents, prefer this schedule format as an effective short-interval scheduling communication tool. But some built environment participants prefer network diagrams as the quote included in the next example from a GC scheduler indicates. Figure 14.2 has added status columns to the previous figure that document actual start and completion dates as well as accommodate forecasting of revised schedule dates. This use of a tabular schedule is also a good as-built schedule.

Example 14.2

"The value of a good network is continually ignored. There is too much focus on the dates in tabular schedules and not enough on understanding the relationships between the activities. You can learn a lot from the graphic that you might otherwise miss when just looking at a tabular schedule."

Mountain Construction Company
Tabular schedule

Project: <u>Fire Station 83</u>   Date: <u>8/30/22</u>
Superintendent: <u>Ralph Henry</u>   Sheet: <u>1 OF 1</u>

No.	Activity description:	Original Scheduled Start date	Forecast or actual Start date	Original Scheduled Finish date	Forecast or actual Finish date
10	SOG finegrade	8/12	A 8/12	8/16	A 8/16
15	Vapor barrier under SOG	8/17	A 8/16	8/18	A 8/17
20	Underslab utilities	8/18	A 8/17	8/22	A 8/23
25	Formwork	8/18	A 8/17	8/23	A 8/21
30	Reinforcement steel	8/23	A 8/19	8/26	A 8/25
35	Place/Pump/Finish 8" SOG	8/26	A 8/26	8/26	A 8/26
40	Place/Pump/Finish 4" SOG	8/29	A 8/29	8/29	A 8/29
45	Spray curing compound	8/29	A 8/26	8/30	A 8/29
50	Remove forms	8/31	A 8/30	8/31	F 8/31
55	Backfill at edge	9/1	F 8/31	9/1	F 8/31
60	Polish SOG	9/2	F 8/31	9/9	F 9/7

Figure 14.2 Tabular schedule update

Pull Schedules

Lean construction includes a variety of skills and techniques including target value design, supply chain material management, off-site prefabrication, just-in-time deliveries, site logistics planning and others as introduced in Chapter 5. The driver for the pull planning schedule is the detailed contract schedule, specifically the critical path necessary to accomplish identified milestones. A pull planning schedule is a version of short-interval scheduling as discussed elsewhere. A photograph of a pull planning schedule from a commercial construction project's meeting room is shown in Figure 14.3. Electronic pull planning software is now available and utilized on larger construction projects, as discussed in Chapter 12.

Concrete Pour Schedules

There are many scheduling tools available to assist the general contractor's superintendent with schedule control. Some of them are developed manually. One schedule control method is through the use of a concrete log that tracks the total concrete expected to be placed on the job compared to how much has been delivered and placed. If 242 cubic yards (CY) of the fire station's total 306 CY of concrete has been placed as of August 10, then the superintendent is comfortable saying he is 79%

Figure 14.3 Pull planning schedule

complete with concrete. This is also an approach for calculating the amount of earthwork imported or exported compared to the total estimate by counting dump trucks that enter and leave the jobsite.

This author believes there is value in some of these manual scheduling methods, including asking the superintendent what percent complete he or she is. They then have buy-in on a variety of schedule control tools, and not complete reliance on a schedule report prepared by the home office.

Equipment Schedule

The general contractor's project superintendent will need to closely track all construction equipment that is on the jobsite. There are several construction management tools they will utilize to accomplish this. On the jobsite, the construction team may utilize a simple Microsoft Excel spreadsheet as an equipment log or equipment schedule that lists each piece of equipment, its source, cost code, arrival date, departure date, and potentially any comments associated with operation and maintenance. The equipment schedule from the case study is shown in Table 14.2. The superintendent's daily job diary is also a good resource for recording equipment arrival, use, and departure as discussed here.

Table 14.2   Equipment schedule

Mountain Construction Company
Fire Station 83
Project equipment schedule

Superintendent: Ralph Henry

Equipment Number	Description	Supplier	Arrival	Rental Rates	Rental Periods	Operator	Scope of Work	Departure	Maintenance Comments
127	Pickup truck	MCC Internal	6/10/22	$800	month	Super.	Misc.		Need oil change
131	Pickup truck	MCC Internal	6/10/22	$800	month	Asst. Super	Misc.		
199	Job office trailer	MCC Internal	7/1/22	$355	month				Need to replace lock
319	Truck crane	Reliable R.	9/9/22	$5,196	month	OE	Steel		
332	40' reach forklift	Reliable R.	7/5/22	$2,600	month	OE	Misc.		Flat tire replaced
371	JLG boomlift	Reliable R.	9/9/22	$3,250	month	IWs/CAs	Steel/Carp		One more wk for CAs
448	Compressor	Reliable R.	7/5/22	$460	week	Carpenters	R.Carp		Watch hoses!
488	225A welder	Reliable R.	9/9/22	$457	month	IWs	Steel	10/21/22	
501	Torch	MCC Internal	9/9/22	$375	month	IWs	Steel	10/21/22	
991, 992	Con-crete vibrators	C. Specialties	7/10/22	$75	day	Laborers	Concr.	9/1/22	2 replaced

Tower Crane Schedules

Each contractor on a project feels they have priority and free use of a tower crane (TC) if a general contractor provides one for the project. This is not the case. The GC's project superintendent will schedule the crane's use by giving priority to (A) critical path activities, (B) delivery trucks that are blocking the road or material storage impeding productivity, (C) safety issues, and (D) GC direct work tasks will take precedence over subcontractor tasks. It is not uncommon for subcontractors to schedule their picks after normal working hours or on Saturday and will be responsible for the overtime costs of the operator and rigger and potential additional rent of the crane. The superintendent will prepare a simple spreadsheet schedule for TC use and present it at the Monday morning foreman's meeting. These are all valuable schedule communication tools. It would behoove a subcontractor to review TC accessibility rules before submitting a bid to a GC or signing a subcontract agreement.

Daily Job Diary

The daily job diary is an important construction management document and scheduling tool. Some construction professionals refer to the diary as a daily journal, daily report, or daily log, and other hybrids. This daily report is different from the formal reporting tools discussed in the next section. The name is not as important as the fact that the document is used properly. The superintendent's job diary is one of the most important construction management tools and jobsite scheduling records. It is viewed as a reliable record of each day's events. Its uses include:

- Contemporaneous and therefore often legal record of daily jobsite activity;
- Change order backup documentation;
- Back charge and claim documentation;
- Historical record of schedule progress, and others.

An example of a job diary for the case study project is shown in Figure 14.4. There are a variety of communication forms that are utilized throughout the construction industry. Each construction firm should select a standard format to use on all projects. Following is a list of the important items to be included with any construction diary:

- Superintendent's name,
- Job name and reference number,
- Date and day the diary is filled out,
- Weather,
- Manpower for both the GC and subcontractors,
- Deliveries received,
- Work accomplished,

Mountain Construction Company
Daily job diary

Project: Fire Station 83   Job number: 9922
Superintendent: Ralph Henry   Date: 10/17/22   Day: Monday
Today's weather: Overcast in the am but burned off, 10MPH wind

Activites completed: A few connections left for trusses, removed bridging. Metal deck about 15% down, Wood framing continuing, Plumber and electrician following framer with drilling and setting boxes. We had our 8am safety meeting, 12 guys here, all signed in.

Problems encountered: There were some prefab errors with shear walls which required fixing. Electricians were pulled off the job at 1p – we need at least one here all day long in case we have temp power issues.

Materials and equipment received today: Replacement nailgun arrived, Plumber received a load of copper and plastic, Plywood sheeting arrived – ready for install later this week.

Rental equipment returned: Welder which busted last week was finally picked up by Reliable Rentals – They agreed not to charge rent for last week.

On-site labor/crafts/subcontractors/hours worked:
    Carpenters: 6 @ 8 hours/ea
    Laborers: 1, cleanup only
    Ironworkers: 1 for a couple of hours to pick up tools
    CM Finishers: 0
    Masons: 0
    Electricians: 2 until 1pm
    Drywallers: 0
    Painters: 0
    Plumbers: 2 for 8+ hours, arrived early ~ 7am
    HVAC Tinners: Foreman swung by mid day to scope work out for next
                 week
    Other: City inspector came by at 10am to check out trusses, shear wall
               issue, and wood framing progress. Warned electrician not to
               install conduit or wire until roof is on.

Figure 14.4   Daily job diary

- Hindrance to normal progress,
- Inspections,
- Visitors to the site,
- Equipment on site,
- Accidents, and
- Routing or distribution of copies.

It is also a good practice to attach construction photographs to the daily diary to illustrate job conditions each day. The Fire Station 83 specification section 013200.2.4 lists 15 items the owner expects to be included in MCC's superintendent's daily report; most of them were listed here.

The project superintendent is responsible to prepare the daily diary. Even if the diary is delegated to an assistant superintendent, foreman, or field engineer, the project superintendent must take ownership and approve before it is forwarded to other corporate team members. But because of the historical and potentially legal importance of the diary, it should be prepared by the project superintendent, who is responsible for all field activities. The diary should be completed daily, at the end of the day. Postponing completing the diary until the next morning or assembling five of them at the end of the week dilutes their accuracy.

It is a good practice to require onsite subcontractors to also complete daily diaries and provide copies to the general contractor's project manager and superintendent. This way the project superintendent has the original evidence of the subcontractor's manpower and their view of daily progress. If there is a problem or restraint noted on a subcontractor's diary, the GC's project team should deal with it immediately. Some project owners also may require copies of the GC's superintendent's diary as well as from major subcontractors, especially negotiated or design-build subcontractors. There are many other specialty schedules that are valuable superintendent tools, including inspection schedules and punch list schedules.

14.5 REPORTS

Contractors prepare many reports and review and act on reports prepared by others. Many construction reports include schedule data. These reports may be electronic or on paper. If electronic, because of their significance as a record document, most end up printed and a copy will be filed. The superintendent's daily job diary discussed above may also be called the daily report. The reports discussed in this section are likely monthly executive reports that are shared with upper management and/or the project owner, which are different from the superintendent's daily report.

One of the most common internal reports prepared by the PM is the monthly cost and fee forecast as introduced earlier. This document may be expanded into a corporate reporting tool and include updates on schedule, safety, QC, and client relations, in addition to cost and fee reporting. Many open-book negotiated projects require the GC's PM to prepare a monthly executive report for the client's use. This report will also status cost, schedule, quality, and safety updates.

There are a multitude of scheduling reports created by scheduling software. These can be created as often as desired by the press of a button, as was discussed in Chapter 12 with schedule technology. The software systems can create lists of detailed or summary information regarding schedule status, such as:

- 105 activities were started since the last report update (and a link is provided to list out each of those activities);
- 110 activities were completed since the last report update;
- 1,021 activities are in progress;
  - Of these, 274 activities are on schedule,
  - 399 are ahead of schedule by one day or more,
  - 348 are behind schedule,
- Of these, 107 activities are on the critical path. As a result, the current contractual critical path has seven days of negative float.

Reports often find their way into claim preparation and/or claim defense therefore they must be accurate and objective. Reports should be printed, edited, receive a third-party review, and repeat the QC report development process. Good examples of this QC process are reflected below. A hard copy of every revision, which is published and distributed, should be kept, similar to record drawings. Similarly, contractors must print and retain hard copies of all formal reports that are distributed

Similar to schedules, reports are construction communication tools. It is important for the author to know his or her audience with respect to status reports and schedules. Reports that document everything, such as the status of each of the 10,000 activities on a large public works project, may be too large and get out of hand. The recipient may not open it or may miss the main point. Summary or executive reports eliminate a lot of the detail and status only the big-picture issues. In this way, the reader does not get "lost in the weeds."

Example 14.3

Many clients (and contractors) request monthly executive reports outlining project status related to cost, safety, quality, and schedule concerns, among others. This particular project required these reports to eventually be forwarded to a board of directors which were not necessarily versed on all construction terminology. The GC's PM would run a draft of the monthly report by the owner's representative before finalizing and printing a dozen color bound copies. There would still be questions, but at least the owner's representative was prepared for them and had buy-in on the report.

Example 14.4

A construction management consultant was hired to perform a three-month audit on a school district's large multischool expansion. The school board asked for a formal report once complete, but the superintendent of schools was reluctant to guide the consultant on the format or content of the report, other than to say "We live in a fish bowl here." This implied the report needed to be sensitive, but at the same time, the consultant was bound to be accurate and complete. He prepared three reports: a rough draft red copy, a second draft yellow copy, and a final green copy. The consultant and the superintendent reviewed the red and yellow copies together and achieved buy-in before the final green version was bound and successfully distributed to the board members.

There are potential problems with all electronically transmitted reports, such as the sender indicating "Well, I sent you that report via email," but the recipient replying "I didn't open and read it all." Because of this, many managers will send a hard copy in addition to an electronic copy. Depending on the content of the report it may also be sent via certified mail and comply with notice requirements described in the contract agreement.

14.6 TECHNOLOGY TOOLS

The computer is a useful tool in preparing a multitude of schedule reports. Reports can be an efficient means of communication to various stakeholders. Not all involved in a construction project can read a detailed schedule but some may find lists of data, especially on a summary or milestone level, more understandable. The reports represent a snapshot in time and do not always tell the entire picture. It is important that they be accurate, as reports are often utilized as backup for claim preparation as will be discussed in Chapter 18.

It is important for all built environment team members not to let the computer be the scheduler; the scheduler must think about the input and results. This is a risk or potential problem if the computer auto updates without the input from the jobsite management team. The computer can very easily generate reports by just the push of a button, but the computer does not necessarily "think" about the report's content or presentation. It is important that the scheduler review and QC a computer-generated scheduling report before sharing with other team members, both internally and externally.

A forecast for completion, provided as part of a schedule status or update can be accomplished with software as well, but the original date should be maintained as long as possible. One critique of revised schedules is the paper trail, documenting "How did we get here?" Scheduling software tracking tools allow retention of the original schedule. Color-coded schedule activity updates work on the computer screen, but color copies can be expensive. Black-and-white copies of a color schedule will not show the color codes. Software can simply and effectively use two bars to show the original schedule and current progress, if it is desired that a new schedule document be produced.

Various scheduling software programs allow production of schedule tools such as a *schedule exception report*. This is a listing of activities sorted by:

- Activities started,
- Activities completed,
- Actual durations versus scheduled,
- Manpower spent versus planned,
- Materials delivered, and other sorts.

These reports can be helpful in schedule control efforts, but because of their ease of production, they can also be overwhelming. It is important that the reports do not take on a life of their own. The scheduler should remember that the ultimate goal of the schedule is as a tool to assist completion of the project timely – not report on reports.

Building Information Modeling

Building information modeling (BIM) plays an important role in preconstruction planning as was discussed in Chapter 3. BIM has replaced many two-dimension drawings (2D) with three-dimension drawings (3D) and then added a fourth-dimension (4D) – time – and now cost has been added as a fifth dimension (5D).

Often the subcontractors will utilize computer-aided design and BIM and place multidisciplines' work on the same drawings. These shop drawings, or installation drawings, are known as coordination drawings. These are intended to resolve conflicts on paper, or in this case, on the computer screen, prior to materials arriving on the jobsite and saving all firms substantial costs in refabrication and rework. Coordination drawings are often a form of submittal, as discussed earlier.

Resources

Some schedulers advocate placing material quantities and construction costs, including equipment and labor, on the detailed construction schedule. But if this is done all in one place, it should not be distributed to other external project stakeholders, such as the project owner and subcontractors, and as stated before, it may be counterproductive to keep two different schedules. Placing cost data on the schedule negates the purpose of a construction schedule as a communication tool. Cost management can best be accomplished through work packages, as discussed in the next chapter. The schedule communicates the project plan to many stakeholders, not just the GC's PM and superintendent. If the schedule is cluttered with too much information, similar to the "horse in the barnyard example" presented in an earlier chapter, it loses its effectiveness as a communication tool for the builders.

14.7 SUMMARY

Submittals are documents or products that are prepared by contractors and suppliers and forwarded to the design team for review. They represent the final phase of the design process, and are key components of the GC's schedule management program. Submittals include shop drawings, manufacturer's technical data, and product samples. They may be prepared by subcontractors, suppliers, or the GC's technical staff. Because of their potential impact on the construction schedule, all submittal requirements should be identified at the beginning of the project and tracked using a submittal schedule. The submittal schedule is used for managing the submittal process with the design team, and an expediting log is used for managing the submittal and material processes with subcontractors and suppliers. An effectively managed submittal program is a necessary tool to achieve a successful construction project schedule. Required material delivery dates must support project schedule requirements.

Project management tools which are directly related to the detailed construction schedule include the monthly pay request and monthly cost and fee forecast. Float management is also a key focus of construction PMs. Recording what actually occurs on a project with respect to deliveries, start and completion dates, manpower, and others on the detailed schedule produces a valuable as-built schedule.

Superintendent scheduling tools include short-interval schedules, tabular schedules, pull schedules, and others. Manual logs or schedules are also tools for superintendents and include concrete and earthwork schedules, equipment schedules, and tower crane schedules. The superintendent's daily job diary also reports schedule progress and is an important historical recording tool.

A general construction project manager often prepares monthly or quarterly executive reports for his or her home office management team and potentially for the project owner on a negotiated project. These reports include summaries of project progress for both time and cost as well as summaries of safety and quality controls. Schedules are useful construction communication tools if planned, prepared, and incorporated into the daily work plan properly. If the schedule is used as a tool to punish those who miss a date here or there, its effectiveness will be compromised as discussed in the next example and a few others in this book.

Example 14.5

If cost engineers and schedule engineers are not integral and part of design or construction teams, but rather are there to report progress or delays, then trust does not exist between the parties. In order for schedulers to get realistic information from those that are performing the work, the designers and builders need to be confident that they will not be punished with that information later. The lead scheduler on this large industrial project would task his area schedulers to essentially find and report to him where the job was behind schedule, so that the lead could take the information 'up the ladder' to his superiors in hopes of a reward. The area schedulers should have been on the job to assist the designers and builders, providing schedules as tools to help with the tasks at hand, rather than using them as a negative reporting tool.

14.8 REVIEW QUESTIONS

1. Who should prepare daily job diaries and why him or her?
2. Who should participate in pull planning schedule development and why them?
3. What are some methods proposed in this chapter to ensure that only quality schedule reports are produced?
4. What is the difference between a fabrication drawing, erection drawing, and coordination drawing? Which of these is considered a submittal or shop drawing?
5. Who on the GC's team takes the lead in preparation of the pay request and monthly fee forecast?

14.9 EXERCISES

1. When do subcontractors typically get to use the GC's TC?
2. What percent complete is the fire station case study project based on the SOV included on the companion website?
3. What percent complete is the fire station's SOG based upon the tabular schedule presented in Figure 14.2?
4. Why might a GC not share its expediting log (Chapter 17) with a project owner and designer?
5. Assume that a new replacement carpet discussed in Example 14.1 is 12 weeks out. The furniture is coming off the truck and is ready for installation. The owner needs to move new people in this week. What should the team do? Who pays for any rework? Did the client find their new GC team member? Should contractors submit only what is asked? Why would the design team not request that everything be submitted? What happens if a contractor submits on a product for which a submittal was not requested? Will the design team review it? Will the architect's errors and omissions insurance pay for removal and replacement and impact costs if this carpet is returned? Does the subcontractor have any liability?

C h a p t e r **15**

Jobsite Control Systems

15.1 INTRODUCTION

As stated prior, project control systems for both the project manager (PM) and superintendent include:

- Safety control;
- Cost control, including change orders and pay requests;
- Quality control (QC);
- Document control; and
- Schedule control.

This book is of course primarily focused on just one of these control aspects, schedule control, which is discussed in detail in Chapter 13. But any book on construction management tools would be amiss not to include a brief discussion on these other important control aspects. This chapter addresses safety control, cost control, and quality control, which includes inspections and punch lists. Document control is threaded throughout the book. These five areas of control must work together homogeneously for the jobsite team to achieve success as reflected in Figure 15.1. If any one of these areas fails, the project will fail. Schedule control is depicted in the center of this figure because that is the focus of the book, but it does not necessarily take precedence over the other four control systems. In addition, several other jobsite management tools or processes are described here and elsewhere, including:

- Environmental controls,
- Traffic control,
- Jobsite laydown management (also see Chapter 5),
- Material management (also see Chapter 17),
- Equipment management (also see Chapters 10 and 14),
- Labor relations and productivity (Chapter 10),
- Cash flow management (Chapter 11),
- Subcontract management (Chapter 17),
- Risk management (Chapters 6 and 18), and others.

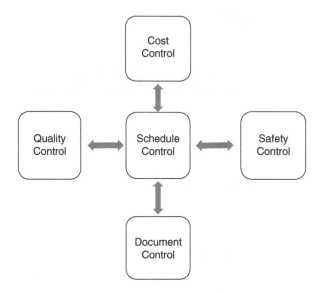

Figure 15.1   Construction management control systems

The preconstruction plans that are introduced in Chapter 3 require implementation and control during construction. Control involves measuring success by the project team against those plans. If the results are positive, it reflects that the project is proceeding as planned. If not, modifications are necessary to the planned process. Many of these control systems could warrant their own separate chapters, as is done with many construction project management textbooks, such as *Management of Construction Projects, A Constructor's Perspective.*

Construction participants utilize the term *project controls*, but can construction managers actually "control" people or a company or a situation? Can a laborer be forced to shovel 20 cubic yards (CYs) of dirt in one day? Can a general contractor (GC) force a drywall subcontractor to finish the elevator shaft before the inspector arrives on Tuesday? Field managers really cannot control, but what they can do is "manage," and that is done by placing the right craftsmen and contractors with the right tools and materials and accurate drawings together. Project managers and superintendents support their team by providing them with the resources they need to be successful. For the sake of this conversation, the two terms *control* and *manage* will be intermixed, as is done in the industry, in most textbooks, and in the classroom. But going forward, the next generation of construction managers should focus on honing their communication and team-assembly skills such that schedule, safety, budget, quality, labor, materials, equipment, and subcontractors are all "managed" efficiently.

15.2 SAFETY CONTROL

Safety control, like project management and scheduling and estimating, warrants its own book and class as well. Safety control is important for a multitude of reasons including the fact that accidents happen. Safety infractions cost money and delay the schedule, and employee morale is lower on projects that are not built in a safe fashion. The GC's project superintendent should have an in-depth knowledge of the Occupational Safety and Health Administration (OSHA) requirements as well as his or her corporate safety policies. The superintendent will collect safety data sheets from all subcontractors and suppliers, maybe with help from a field engineer, and keep these records on file and accessible for all to review. A large three-ring tabbed binder is often the most efficient document control tool in this case. Each jobsite must have a designated competent safety person. This individual is trained and authorized in safety planning, controls and implementation. For smaller projects, this is the project superintendent. It should not be the corporate safety officer who visits the site once weekly and who is not in a position to daily hold craftsmen and subcontractors accountable when they are off-site. Larger projects will have a full-time safety manager assigned to the project who can serve as the designated person. The designated person must be trained to deal with OSHA inspections and inspectors and associated documentation.

Direct labor costs contractors much more than just the wages the craftsmen and supervisors receive on their paycheck. Contractors pay an additional markup, or percentage add-on, on top of all of the wages they pay; this is known as labor burden. Labor burden is not a fee or profit markup, but it is a direct cost of doing work. Workers' compensation insurance markup, or workers' comp, is a major element of labor burden and varies considerably due to a variety of factors including the safety record of the contractor and its associated experience modification rate (EMR) and the potential safety risk of the labor craft. The baseline EMR is 1.0. Contractors which have a higher incident-rate of safety accidents have an EMR rate greater than 1.0 and those with fewer accidents a rate below 1.0. Some crafts are more prone to accidents, and they will have a higher worker's comp rate. A higher EMR rate increases worker's compensation insurance rates which make it more difficult for the contractor to be a successful low bidder.

The general contractor's project superintendent is the individual ultimately responsible for safety control. Since subcontractors often make up 80–90% of the work force, the focus on safety needs to include subcontractors. Some project superintendents will argue that it is actually more difficult to control safety with subcontractors than with their own direct crews. One good method of safety control is an orientation meeting for every subcontractor and each of its employees when they first mobilize onto the jobsite.

15.3 COST CONTROL

All of the construction control systems are critical, but because the goal of all contractors is to make a profit, cost control receives a lot of contractor attention. Cost control is also discussed in most construction management and project management textbooks, including *Cost Accounting and Financial Management for Construction Project Managers*. Much of the material in this section has been borrowed from that source and focuses on the cost-control responsibilities of construction field managers.

Cost control begins with a good construction estimate and detailed schedule. These tools need to be as accurate as possible to allow the project team a reasonable opportunity to meet or beat the company's fee goals. The major elements in any construction estimate include: direct labor, direct material, construction equipment, subcontractors, jobsite general conditions, home office overhead, and other percentage markups including profit.

The markups, including home office overhead, are beyond the control of the jobsite team and are the focus of the chief executive officer and chief financial officer. If subcontractors and suppliers were bought out diligently, and tight agreements written, then there is little to control with respect to their costs. Direct materials should have been accurately measured and competitive prices applied to those quantities. The superintendent cannot install less concrete in the footings than was shown in the structural drawings to save money, so cost exposure is limited for materials as well. Most of the items within the jobsite general conditions estimate can be managed, but the biggest variable there is time – if the project lasts longer, jobsite general conditions will run over. Construction equipment should be cost-coded to direct work and not general conditions wherever possible; this is a premise of activity-based costing. Efficient cost and time management of equipment is necessary. *Direct labor is the most difficult variable* for the estimator to forecast and is the most difficult for foremen and superintendents to control, and, therefore, deserves the majority of focus of any discussion on construction cost control. Direct labor is a risk, but it also provides the greatest opportunity to save money and improve fee.

Similar to schedule control presented earlier, there are five phases of cost control in a typical contractor's cost-control cycle. These include *setup* or estimate preparation, *correct* or adjust the estimate after buyout and input to the cost-control and accounting system, *record* or monitor costs including cost coding, *modify* the system if costs are not being achieved, and preparation of an *as-built* estimate and input back into the company database for use in the next estimate.

The key to getting foremen and superintendents involved in cost control is to get their personal commitment to the process. One successful way for the GC to do this is to have the PM and superintendent actively involved in developing the original estimate as previously discussed. If the superintendent said it will take two carpenters working three days to form the spot footings, he or she will endeavor to see to it that the task is completed within that time.

One very broad technique for monitoring project cost is to develop a *whole project direct work labor curve*. In this approach all of the direct work is combined into one curve; it is not separated by craft or work package as discussed below. It is important to have the superintendent or foremen record the actual hours incurred weekly and chart them against the estimate; the jobsite cost engineer or scheduler can help with this. If the actual labor used is under the curve, it is assumed the foremen and their crews are either beating the estimate or behind schedule. The opposite is true if the hours are above the curve. Direct labor is an important scheduling resource and this curve is introduced in Chapter 10. This simple method of recording the man-hours provides immediate and positive feedback to the project team, but it has its limitations. It is better to use hours and not dollars when monitoring direct labor. This is one advantage of estimating with man-hours over unit prices for labor. Foremen and superintendents think in terms of crew size and duration. They do not think in terms of $500 per opening to install doors, frames, and hardware; rather, they consider that they have scheduled two carpenters to work six hours on the task.

Preparation of the original estimate, schedule, and supporting work breakdown structure are done more efficiently by *work packages* or assemblies than by using the MasterFormat divisions developed by the Construction Specification Institute. Work packages are a method of breaking down the estimate into distinct packages or assemblies or systems that match measurable work activities. For example, concrete footings, including formwork, reinforcing steel, and concrete placement could be a work package. When the footings are complete, the superintendent and the management team will have immediate cost and schedule control feedback. Each work package should be developed by the foreman who is responsible for accomplishing the work, maybe with the assistance from the field engineer. Figure 15.2 shows an example of a foreman's cost-control work package for the case study's concrete stem walls.

Some construction executives believe that field supervisors should not be told the true budgeted value and duration of each work package. This is a poor practice, because the foremen and superintendents are key members of the project team and have critical roles in achieving project financial success. These men and women are regarded as the last planners and should be provided the actual budgeted cost for installation, both in materials and man-hours, as well as the scheduled time for installation for each work package.

Recording hours is not the same thing as controlling or even monitoring, but recording leads into control. Cost and time recording begin with accurate but workable activity codes. If the cost and time are not recorded accurately, then the

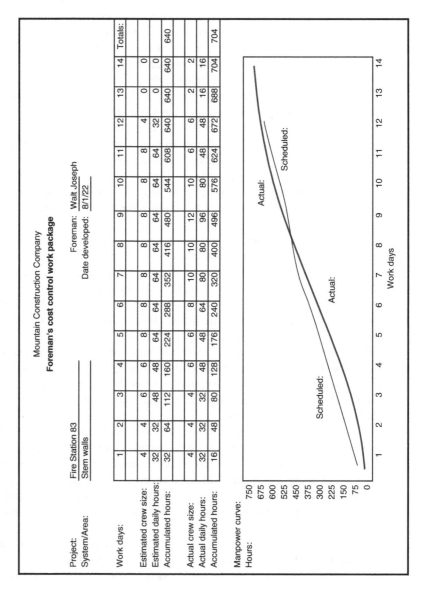

Figure 15.2 Foreman's cost control work package

information collected is not usable for future projects. The next two examples were provided by experienced cost engineers, but the same concepts apply to schedulers.

Change Order Processing

Change order and pay request management both have serious impacts on a contractor's ability to maintain positive cash flow. Timing is crucial with respect to change order notifications. If the superintendent discovers unknown site conditions, but does not notify the project owner within a contractually dictated time period, such as 21 days, which is the standard for American Institute of Architects A201 general conditions, the contractor may have waived its right to prepare a claim. It is important that contract clauses like this be discussed between the home office, PM, and superintendent at their internal preconstruction meeting. It is also important for the superintendent to be aware of how extra work is assigned to the contractor by the project owner and design team and respond accordingly – they cannot just casually ask him or her to perform extra

Example 15.1

This cost engineer was required to review timesheets daily for 1,100 pipe fitters on a very large power-plant project. One particular pipe hanger seemed to have 300 or so man-hours charged to it daily. On further inspection by the cost engineer, that hanger was discovered right above the door exiting the turbine building. The pipe and hanger had been there for years and had cobwebs on them. But what was very clear for all of the foremen looking for an easy timesheet cost code as they finished their work day and exited the building was this one pipe hanger number.

Example 15.2

This cost engineer was responsible for tracking and reporting labor costs on a large hydroelectric project. Her focus area was for the electrician and plumber trades, which amounted to 600 and 300 employees respectively. At the end of every shift, foremen would write, in hand, often with hurried handwriting, the work each individual craftsman performed that day, along with their hours. A separate timekeeper working for payroll would translate these timesheets into cost codes, which would be entered into the accounts payable system each evening. The next morning the cost engineer would be presented with an Exceptions Report, which would kick out all of the cost-coding errors. She would spend the first half of each day tracking down the foreman and timekeeper and inspecting the work on the site to rectify these miss-codes and then journal entry them into correct codes. And the next day's activities would be a repeat of this one. This is an example of cost reporting, not cost control, and the same premises are involved with schedule control.

work; it must be documented in writing. Significant change orders may turn into claims or disputes and can also impact the construction schedule, as discussed in Chapter 18.

Pay Request Processing

Schedule updates are often required by project owners to accompany the GC's monthly pay request. In preparation of the monthly pay request, the general contractor's project manager and superintendent will:

- Review subcontractor and supplier pay request proposals for percentage completion.
- Walk the site to verify if materials have been delivered and work has been accomplished.
- Initial invoice approvals, which is a method of financial checks and balances.

The best method for a contractor to facilitate expeditious change order and pay request processing is to be ethical, honest, and fair with their pricing. Prices should not be inflated or front-end-loaded. Contractors should be prepared to share open-book change order pricing and the pay request schedule of values (SOV) with the project owner and/or its accountant or bank, even if it is a lump sum project. The SOV is a direct byproduct of both the estimate and schedule as was discussed in the last chapter. The fire station's pay request SOV is on the book's companion website. The contractor's financial accounting books should be prepared such that they are always ready for an owner to audit them. Timing of payment receipt has a stepped impact to the cash flow schedule, as discussed in Chapter 11.

15.4 QUALITY CONTROL

Quality control (QC), or quality management, is also an important project management function and is one of the five critical attributes of project success, with the others being cost, time, safety, and documentation. QC has short-term implications affecting material and labor costs on a project and long-term implications affecting the overall reputation of the construction firm. Nonconforming materials and work must be replaced at the contractor's cost, in terms of both time and money. This means that the contractor must absorb the financial cost of tearing out and replacing the nonconforming work and additional contract time is not granted for the impact the rework has on the overall construction schedule. Quality management is essential to ensure all contract requirements are achieved with a minimum of rework. This requires a proactive quality focus from all members of the project team.

To achieve quality in the project, the project manager and superintendent must ensure management systems are in place to assure that quality materials are procured and received, quality craftsmen and subcontractors are selected, and all workmanship meets or exceeds contract requirements. This proactive approach involves careful planning and an effective inspection process to ensure quality results. Variations from standards are measured, and actions taken to correct all nonconformities, or deficiencies. Effective QC planning starts with a detailed study of the contract documents to determine all of the project's unique quality and testing requirements. Qualified inspectors are selected to ensure that all self-performed and subcontracted work meets contract specifications. These inspectors also should participate in the submittal process to ensure that quality materials are being submitted and that only approved materials are installed. The use of special third-party inspections is also a valuable active QC approach. These inspectors are often directly contracted by the project owner. Mock-ups, which are stand-alone samples of completed work, should be planned and included in the submittal schedule to establish workmanship standards for critical architectural features.

Subcontractors perform most of the work on a construction project. The GC's reputation for quality work, therefore, is greatly affected by the quality of each subcontractor's work. Quality subcontractors are not necessarily the least-cost subcontractors but, in the long term, those that provide a superior project. The term *best-value subcontractors* applies here as well. The superintendent should walk the project when each subcontractor finishes its portion of the work to identify any needed rework. Any deficiencies noted should be listed in the QC log or punch list. All rework needs to be completed before follow-on subcontractors are allowed to start work. Deficiencies should be corrected as the work progresses to minimize the size of the punch list during project close-out. If punch lists are excessively long and not corrected timely, substantial completion will be impacted and receipt of final retention withheld.

15.5 DOCUMENT CONTROL

Communication skills are one of the most essential construction leadership skills. Construction field supervisors who know the work but are unable to communicate their plans to other project team members will have difficulty succeeding at the superintendent or PM level. Many of the means of communication in construction still rely on paper documentation, although much of that is moving to electronic communication today. Communication tools are more than just written documents. Listening is perhaps the most crucial communication tool, and good listening skills are always among the top of any list of construction leadership skills.

Some of the important construction document control tools that field managers must master include the daily job diary, meetings and meeting notes, requests for information, and submittals. These and other document examples involving PMs and superintendents are discussed throughout this book and include different types of schedules and various logs. The contract schedule itself a 'document,' even if it only exists electronically.

15.6 ADDITIONAL JOBSITE CONTROL SYSTEMS

All plans developed by the GC during the preconstruction phase also require implementation and control by the project team. If actual conditions differ from the plan, adjustments are necessary. Most of these additional focus areas are the responsibility of the GC's project superintendent. A more in-depth description of the role and responsibilities of superintendents is included in *Construction Superintendents, Essential Skills for the Next Generation,* and the interested reader should look to a resource such as this.

Environmental Controls

There may be multiple environmental restrictions placed on construction operations, depending upon the location of the construction site. There may be noise restrictions at night or the site may be near a protected body of water or wetlands. Environmental planning includes dust control, pollution control, noise control, stormwater control, traffic planning including truck sizes and load weight limits, construction parking, concrete truck-wash areas, street sweeping, demolition, and waste removal. Some of the typical environmental control activities include silt fence, catch basin protection, protection of significant trees, stormwater ponds or tanks, and others. Environmental planning also includes removal of hazardous materials pre-existing on the site before construction commences, as well as because of the construction process. It takes time for the superintendent to implement environmental controls, but it will take more time if not planned properly or a mishap occurs. Implementation of the original environmental control systems should be included in the detailed schedule, often in conjunction with mobilization activities.

Traffic Control

Submitting a traffic plan is a city requirement and will be a condition of the building permit. This traffic plan will be reviewed in detail at the preconstruction meeting with the city. The plan itself includes drawings and narratives with days and times of day when deliveries are allowed by the city. The drawing or map will track delivery trucks from when they leave the freeway until they arrive at the jobsite, including all arterials and side streets. Left and right turns will be shown and require approval from the city. Earthwork trucks and concrete trucks will receive special attention. Methods to unload trucks at the site and staging locations (e.g. for the concrete pump) will also need to be presented. This also connects with the contractor's hoisting plan. Another drawing for trucks leaving from the jobsite and merging back on the freeway will also be required. The use of flaggers and special signage will all be detailed as part of the traffic plan. Development of the traffic plan and receipt of city approval should be shown on the GC's preconstruction schedule.

Jobsite Laydown Management

More than just a plan or proposal or consideration, site logistics planning involves creation of a document, or several documents. The first step in preparing the plan is to start with a basic site drawing. Most commercial projects have a civil site plan, a landscape site plan, and an architectural site plan. In developing the jobsite layout plan, the superintendent should consider site constraints, equipment constraints, jobsite productivity, material handling, and safety. In addition to its use for labor productivity, the jobsite layout plan is a great proposal/interview/marketing tool. It shows the owner that the contractor has thought through the project – a personal touch. It may make a difference on a close award decision with a negotiated project. Although developed by the project superintendent, it should consider the needs and requirements of all subcontractors working on site, not just the GC's needs.

Efficient jobsite organization is essential for a productive construction project. The jobsite layout affects the cost of material handling, labor productivity, and the use of major equipment by the general contractor and the subcontractors. A well-organized site has a positive effect on the productivity of the entire project workforce as well as their safety. The jobsite layout plan should identify locations for temporary facilities, material movement, material storage, and material handling equipment. The double-handling of material is wasted effort and not lean construction practices, as stated in Chapter 5. The GC's construction schedule will not include time allowances for inefficient material handling. A jobsite layout plan for the fire station case study is included on the companion website.

Material Management

Before field supervisors order material, or send their pickup truck or jobsite flatbed truck to pick the materials up from the supplier's yard, they should make sure that the project manager had not already negotiated material delivery into the supplier's scope. The supplier would be happy to allow the superintendent to deliver its materials for them. Materials are rarely ordered direct from the original estimate, especially those materials purchased by the project superintendent with short-form purchase orders (POs). Even material quantities from a bill-of-materials list should be re-checked at the jobsite level. Long-form POs are utilized for larger value material purchases, such as those over $1,000 and/or those with long lead times and requiring submittals. Long-form POs are typically prepared by the PM.

Material ordering is also a means of lean construction. Just enough material should be ordered, but not so much that there is leftover or waste. Surplus construction materials, such as framing lumber and plywood, have a way of disappearing from a construction jobsite – that is why many of the craftsmen drive pickup trucks to work! A shortfall of materials can have a detrimental effect on the schedule. Imagine the concrete crew waiting for two additional cubic yards on a 1,000 CY pour, or the framing crew short one wood truss while an apprentice returns to the fabricator with the flatbed truck. Management of material suppliers is also discussed in Chapter 17, "Subcontract Management." Commitment dates for delivery of all major materials, such as structural steel, glazing, electrical switchgear, and others will be shown on the contractor's detailed schedule.

Equipment Management

The superintendent's management of equipment on the project begins with his or her very valuable initial estimate input. For example, if the staff estimator assumed the crew could reach the ceiling space above the fire station's apparatus bay with a scissor lift to install process piping, but the superintendent chose a rolling scaffold and had to plank over the trench drains, the estimate would be short. Conversely if the estimator assumed a small boom truck would be necessary to hoist roofing materials but the superintendent negotiated the supplier to directly hoist the materials from their own delivery truck, they may have overestimated and would not have been low bidder.

The project superintendent will need to closely track all equipment which is on the jobsite. There are several construction management tools they will utilize to accomplish this. On the jobsite, the construction team may utilize a simple Microsoft Excel spreadsheet as an equipment schedule, which lists each piece of equipment, its source, cost code, arrival date, departure date, and potentially any comments associated with operation and maintenance. Construction equipment must be kept in good operational condition. The breakdown of key equipment, such as a troweling machine, forklift, or tower crane, can impact the schedule. The superintendent's daily job diary is also a good resource for recording equipment arrival, use, and departure. Examples of both of these scheduling tools are included in the previous chapter.

15.7 SUMMARY

In addition to schedule control, the project superintendent and project manager work together to manage a variety of controls, including safety, quality, and cost controls. There are many tools utilized for each of these processes including preconstruction project-specific safety plans from both the general contractor and all of the subcontractors and weekly safety meetings. Cost control for a contractor often focuses on the area that is most difficult to estimate and control: direct labor. Labor hours are utilized for project labor curves and work packages as efficient jobsite cost-control techniques. In the next chapter a third earned value curve is explained as an even more effective cost-control tool. Expeditious approval of pay requests and change orders and receipt of timely payments facilitates a positive cash flow which is essential for all GCs and subcontractors.

An effective quality management program is essential to project success, both in terms of a satisfied owner and a profitable project for the general contractor. Poor quality work costs the contractor both time and money and can cause the loss of future projects from the owner. The PM and superintendent must ensure that all materials and work conform to contract requirements. The objective of a quality management program is to achieve required quality standards with a minimum of rework. Quality materials must be procured, and qualified craftsmen selected to install them. Workmanship must meet or exceed contractual requirements.

In construction, most think of four primary control systems or focuses: Cost, schedule, quality, and safety. However, document control has been added as a fifth area for which project managers and superintendents are responsible. Schedule control was singled out as a priority skill for this book and is covered in several chapters. Several other plans require control or management by the GC's field management team. Some of these plans requiring management and control include environment, traffic, laydown, material deliveries, and equipment use. If these plans are not managed efficiently, they each can seriously impact the contractor's schedule performance and expose the project to liquidated damages or delayed release of retention or both. The superintendent and project manager must balance all of these factors in order to produce a successful construction project.

15.8 REVIEW QUESTIONS

1. Who are the last planners?
2. What are the five areas of project control?
3. Explain the connection between safety accidents, EMR, worker's comp insurance, and successful bids.
4. If a project lasts two months longer than expected, other than direct labor, what portion of a contractor's estimate is affected?
5. How is a foreman's work package a more effective cost-control tool than a whole project labor curve?
6. How did the fire station concrete stem wall work package depicted in Figure 15.2 turn out: on, ahead, or behind schedule and on, under, or over budget?

15.9 EXERCISES

1. Of the five primary areas of project control, which is least important and could be eliminated?
2. Of the five areas of project control, which is the most important and deserves the most focus?
3. In addition to concrete stem walls, what are three different examples of potential cost and schedule control work packages for the case study project?
4. What is retention?
5. "Schedule" was placed in the center of the project controls Figure 15.1 because this is primarily a scheduling book, but not necessarily because it is the most

important control aspect. Make a connection between schedule and each of the other four control areas.

6. What is a "pay when paid" clause?

7. What is front-end loading?

8. List quality control errors from all of the parties discussed in Example 14.1 in the previous chapter.

9. What is the difference between open- versus closed-book with respect to pay requests and change orders?

10. Best-value subcontractors and supplier selection is a foundation of schedule and quality control. What are some of the early active QC steps a contractor should take to assure selection of best-value team members?

11. Utilizing another case study project, prepare a work package for direct labor for a system of work, such as concrete foundations, cast-in-place concrete walls, tilt-up concrete walls, or slab-on-grade. Attach the original portion of the estimate that refers to this work package.

12. As an alternate to a separate case study project, the following quantity and man-hour columns can be used as a baseline estimate. Assume a three-person crew working for six months from start to completion of all of these activities.

	Estimated Quantities	Estimated Hours
Excavation:	200 CY	300
Formwork:	4,500 SF	650
Reinforcement steel or mesh:	50 tons	400
Concrete placement:	600 CY	500
Slab finish:	75,000 SF	550
2 x 6 wall framing:	50 MBF	750

Earned Value Management

16.1 INTRODUCTION

Earned value (EV), also known as the earned value method or management, or earned value analysis, is one technique used by some contractors to determine the estimated value of the work completed to date (or earned value) and compare it with the actual cost of the work completed. The most effective use of earned value to a general contractor's (GC's) jobsite management team is to track direct labor, which represents the construction contractor's greatest cost risk. Man-hours and not labor cost is best used for monitoring direct labor as every project and every crew will have a different mix of apprentices, journeymen, and foremen and associated wage rates.

This chapter will discuss origination of the third curve, the earned value curve. Previous to this discussion there was only the scheduled or estimated curve and the actual cost curve. The earned value curve and process is an advanced measure of cost and schedule progress for contractors, as earned value management integrates cost and schedule controls. There are many formulas and indices associated with a study of EV, and a few of these will be introduced here as well. Earned value can be applied to analyze the project as a whole or each activity individually.

In addition to cost and schedule control, earned value can also be utilized with pay requests. Some sophisticated project owners may choose to pay their contractors on an earned value system rather than an estimated or actual expenditure-based system.

16.2 DEVELOPMENT OF THE THIRD CURVE

The work-package curve discussed in Chapter 15 charted the man-hours planned by the foreman and the actual man-hours used for the fire station case study concrete stem wall installation. On the fifth day the actual man-hour curve was below the estimated curve. Did this necessarily represent that the foreman was under budget? Could he have been behind schedule? The project team could actually be ahead or behind schedule and over or under budget because the work package compared only estimated to actual hours and did not consider the amount of work accomplished.

The addition of a third, or earned value curve combines estimated with actual cost and schedule performance, which allows true productivity measurement. The EV curve plots the quantity of work performed by the foremen and his or her crew against the hours that were originally estimated to install that quantity. This third curve is determined by plotting the total number of man-hours estimated for the work package multiplied by the cumulative percent completed. If 58% of the fire station stem walls have been installed by day 8, then 58% of the 640 estimated hours have been "earned." In other words, 371 hours of the 640 hours estimated has been earned and 371 would be plotted on day 8. When this activity or work package system is complete, 100% of the 640 hours will have been earned as reflected in the next formula. 640 hours was originally estimated and the contractor cannot earn any more or less than this. If 600 hours were actually spent at the completion of this work package, they will have realized a profit and conversely if they spent 770 hours, they will have lost money on this activity. But if the contractor uses only an estimated versus spent comparison, without an earned value analysis, they will not know the ultimate outcome of the work package until it is 100% complete.

Hours earned / % Complete = Hours at completion = Estimated hours
371 / 58% = 640 hours estimated

In a pure material quantity perspective, if there are 31 cubic yards (CY) of stem walls in the project, and 18 have been installed, then 18 CY or 58% of the total 31 CY estimate has been earned, regardless of how much was spent on those 18 yards. With this third curve, an actual measurement of productivity can be made. This method of monitoring will provide more accurate feedback to the project team for appropriate cost and schedule reporting and correction if required.

Quantity installed / Total quantity estimated = % complete
18 CY installed / 31 CY estimated = 58% complete

16.3 EARNED VALUE AS A CONSTRUCTION MANAGEMENT CONTROL TOOL

Two different metrics of measurement are now available using all three curves in concert; they are schedule and cost performance. The questions the third EV curve answers are:

- Is the work package right on schedule, ahead of schedule, or behind schedule and by how many days?
- Is the work package right on budget, over budget, or under budget, and by how many hours?

Table 16.1   Earned value performance matrix

Schedule status:		Ahead of schedule	On schedule	Behind schedule
Cost status:	Under budget	1	2	3
	On budget	4	5	6
	Over budget	7	8	9

There are nine different combinations of potential answers to these two questions, as reflected in Performance Matrix Table 16.1.

Schedule Status

There are a couple of ways to determine the schedule or cost status of a project or work package utilizing earned value. One is through a series of simple mathematical formulas as discussed with indices below. The other is to chart EV and measure it against the planned or estimated performance of the work package or project and the actual expenditures. The schedule status is determined by subtracting the actual time used to perform the work from the time scheduled for the work that has been performed. This is the same as measuring the graphical horizontal distance between the earned value and the estimated curves shown in the earned value Figure 16.1. This work package is statused as of day 8. At this point, the foreman had installed 83.3% of the work package and had earned 220 hours of work. 220 hours' worth of work was not scheduled to be in place until day 11. Therefore, the foreman is approximately three days ahead of schedule (11 days – 8 = 3 days). This work package, turned earned value analysis by the addition of the third curve, is for the structural steel column and beam installation (erection and bolt-up) for the fire station case study project.

Utilizing data points that went into creating the graph, as included in Figure 16.1, a more precise measurement of cost and schedule status utilizing earned value can be found. The schedule status is evaluated using the schedule variance (SV). This variance is determined by subtracting the estimated cost in man-hours from the effort necessary to achieve the earned value.

> Schedule variance = Earned value – Estimated and scheduled value

If this variance returns a negative value, it means the project or activity is behind schedule. If it returns a positive value, it means the project or activity is ahead of schedule.

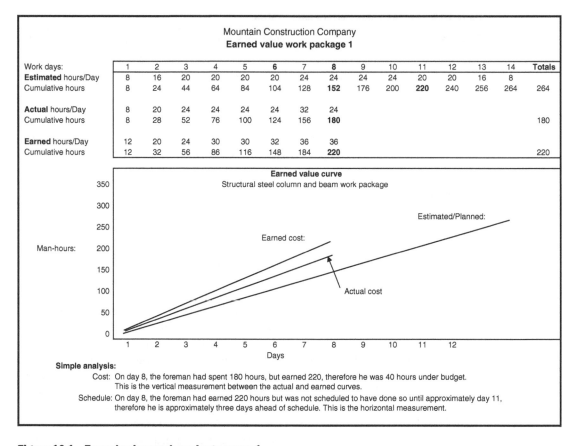

Work days:	1	2	3	4	5	6	7	8	9	10	11	12	13	14	Totals
Estimated hours/Day	8	16	20	20	20	20	24	24	24	24	20	20	16	8	
Cumulative hours	8	24	44	64	84	104	128	**152**	176	200	**220**	240	256	264	264
Actual hours/Day	8	20	24	24	24	24	32	24							
Cumulative hours	8	28	52	76	100	124	156	**180**							180
Earned hours/Day	12	20	24	30	30	32	36	36							
Cumulative hours	12	32	56	86	116	148	184	**220**							220

Mountain Construction Company
Earned value work package 1

Earned value curve
Structural steel column and beam work package

Simple analysis:

Cost: On day 8, the foreman had spent 180 hours, but earned 220, therefore he was 40 hours under budget. This is the vertical measurement between the actual and earned curves.

Schedule: On day 8, the foreman had earned 220 hours but was not scheduled to have done so until approximately day 11, therefore he is approximately three days ahead of schedule. This is the horizontal measurement.

Figure 16.1   Earned value work package curve 1

Cost Status

The cost status is determined by subtracting the actual cost of work performed from the earned value of the work performed. This is the same as measuring the vertical distance between the actual and EV curves. Looking again at Figure 16.1, the actual cost of work performed was 180 man-hours while the earned value was 220 man-hours. Therefore, the foreman is 40 hours under budget (220 hours – 180 = 40 hours). This places the foreman in option one in the performance matrix Table 16.1; ahead of schedule and under budget, which is the most ideal position. Figure 16.2 for the door, frame, and hardware (DFHW) work package shows a different situation. This foreman is over budget, and approximately two days behind schedule and is in option location nine of the performance matrix in Table 16.1 – the worst position.

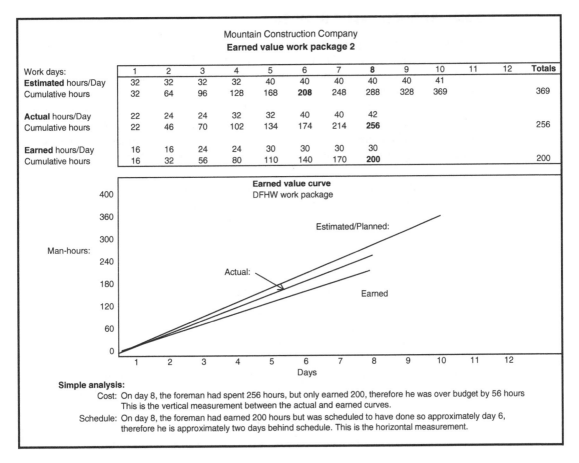

Work days:	1	2	3	4	5	6	7	8	9	10	11	12	Totals
Estimated hours/Day	32	32	32	32	40	40	40	40	40	41			
Cumulative hours	32	64	96	128	168	**208**	248	288	328	369			369
Actual hours/Day	22	24	24	32	32	40	40	42					
Cumulative hours	22	46	70	102	134	174	214	**256**					256
Earned hours/Day	16	16	24	24	30	30	30	30					
Cumulative hours	16	32	56	80	110	140	170	**200**					200

Mountain Construction Company
Earned value work package 2

Earned value curve
DFHW work package

Simple analysis:
Cost: On day 8, the foreman had spent 256 hours, but only earned 200, therefore he was over budget by 56 hours. This is the vertical measurement between the actual and earned curves.
Schedule: On day 8, the foreman had earned 200 hours but was scheduled to have done so approximately day 6, therefore he is approximately two days behind schedule. This is the horizontal measurement.

Figure 16.2   Earned value work package curve 2

Similar to calculating the schedule variance from data points described above, the cost status is evaluated using the cost variance (CV). This variance is determined by subtracting the actual cost in man-hours from the effort necessary to achieve the earned value.

> Cost variance = Earned value – Actual cost

If this variance returns a negative value, it means the project or activity is costing more than estimated. If it returns a positive value, it means the project or activity is costing less effort than estimated.

16.4 EARNED VALUE INDICES

The above curve comparison analysis is a straightforward approach to reviewing the concept of earned value. Another way that many discuss EV is through a series of mathematical formulas and indices. In this case earned and actual curves are not drawn, but points-in-time document the quantity of work installed and the cost of that work, either through man-hours as recommended for cost and schedule control, or through dollars as discussed below with pay requests. These points in time are represented in Figure 16.3. There are a multitude of acronyms and math formulas associated with a technical study of EV; here are just a few of them, along with a layman's explanation.

- ACWP: Actual cost of the work performed. This plots the actual point of hours or dollars when incurred, regardless of what was estimated.
- BCAC: Budgeted cost at completion. This reflects the total value of the work package at the end of the schedule, essentially the top point of the schedule or estimate curve.
- BCWP: Budgeted cost of the work performed. This plots the earned value of work performed and when it was performed, regardless of actual cost.
- BCWS: Budgeted cost of the work scheduled. This is the estimated curve that reflects the original estimate in hours or dollars and when the work was originally scheduled to have been completed.

These acronyms all represent points that could be plotted on the graph. The earned value indices compare these points utilizing the following mathematical calculations:

- Cost performance index, (CPI) = BCWP/ACWP greater than 1 is under budget.
- Cost variance = BCWP – ACWP, greater than zero is within budget.
- Schedule performance index (SPI) = BCWP/BCWS, greater than 1 is ahead of schedule.
- Schedule variance = BCWP – BCWS, greater than zero is ahead of schedule.

Answering the questions of, "is the work package (or project) over or under budget and by how much, and is it ahead or behind schedule and by how much?" at any one specific point in time can now be accomplished mathematically using these indices. The CPI for Figure 16.1 would be 1.2 (220/180), which is greater than one, and therefore under budget. Solving for CV is 40 (220 – 180), which is greater than zero and also under budget. The SPI for Figure 16.1 would be 1.4 (220/152), which is greater than 1 and therefore ahead of schedule. The SV is 72 (220 – 152), which is greater than zero and also reflects that the work package is ahead of schedule. This confirms the approach that was used earlier, which was a simple graphical comparison of horizontal and vertical differences between the three curves.

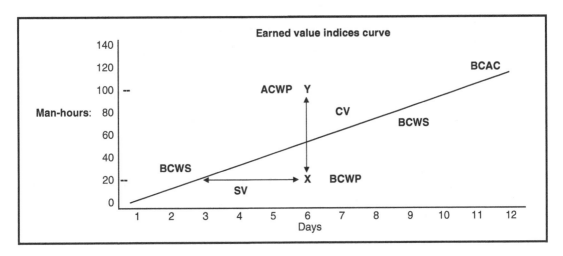

Figure 16.3   Earned value indices curve

Utilizing the data gained from measuring any one point in time can then be extended to forecast what the situation will be at the end of the work package or project. This is similar to cost control forecasting, in which the current productivity trend is assumed to continue until completion. Graphically, this can roughly be done by extending the curves on the same slope as current until the amount of work earned equals the amount of work estimated, regardless of when it is complete or how much is spent. Mathematically, the percentage of the work package or project that is complete, compared to its progress, is used to forecast where the work will be at completion. This is accomplished with the following two formulas.

Forecasted cost variance (FCV) = Cost variance / Percent complete

= Hours or $ over or under at project completion

FCV = 40 HRs / 83.3% = 48 HRs under budget at completion

264 hours estimated – 48 HRs under = 216 HRs total forecast at completion

48 hours savings @ $44/HR = $2,112 savings *

Forecasted schedule variance (FSV) = Schedule variance / % complete

= Days or hours over or under at project completion

FSV = 3 days / 83.3% = 3.6, or ~ 4 days ahead of schedule at completion

14 days scheduled – 4 days ahead = 10 days forecast duration at completion*

---

*Both of these forecast formulas assume a constant trend line (work continuing at the same rate) as a point in time but cost and performance varies so this analysis is somewhat subjective – but it is still better than either a "seat of the pants" forecast or none at all.

Material quantities can be used in lieu of hours or dollars with earned value analysis as well. The letter Q is inserted instead of C in the earlier acronyms, which reflects measurements of quantities scheduled and installed. These acronyms are presented below.

- AQWP: Actual quantity of the work performed
- BQAC: Budgeted quantity of work at completion
- BQWP: Budgeted quantity of the work performed
- BQWS: Budgeted quantity of the work scheduled

Quantity measurements may be based upon percent scheduled and completed. If there are 31 cubic yards of the stem wall concrete estimated at an assembly unit price of $1,477 for a total of $45,772 and 18 yards of concrete are in place, then 58% of the estimated $45,772, or $26,548, has been earned. The alternative is to track exactly how much concrete had been placed. This is sometimes done with a concrete log or by counting concrete trucks and multiplying their load size, such as two trucks times nine cubic yards per truck indicates 18 CY of concrete has been delivered and placed. Due to waste and spillage, this quantity of concrete may be slightly greater than the actual quantity reflected in the formwork.

$$31\,CY\,estimated \times \$1,477\,/\,CY = \$45,772\,estimated$$
$$18\,CY\,installed\,/\,31\,CY\,estimated = 58\,\%\,installed$$
$$58\,\%\,installed \times \$45,772\,estimated = \$25,548\,earned$$

16.5 FORECASTING

The project manager prepares a monthly cost and fee forecast for management that is essentially a new estimate for every line item in the project budget. As mentioned previously in this book, the most difficult element of that effort is forecasting the cost or hours to complete, or to-go, for each activity. There are various methods the PM may use; all may be applicable for different items, but none are applicable for all items. Here are some of the potential choices.

1. Forecasting the final cost to equal the original budget such that the to-go cost equals the difference and the final variance will be zero. This is appropriate if little to no work has been performed.

$$\text{Forecast to go} = \text{Original estimate} - \text{Cost to date}$$

$$\text{Variance} = \text{Forecast} - \text{Original} = \text{Zero cost difference}$$

    a. If substantial amounts of work have been performed, and actual productivity is beating the original estimate, this is an overly conservative approach.

    b. If substantial amounts of work have been performed, and actual productivity is poorer than the original estimate, this is an overly optimistic approach.

2. Forecasting the original productivity rate multiplied times the quantity to-go to equals the cost or hours to-go. This assumes the original productivity rate was more applicable than the trending productivity rate for the balance of the work to complete.

$$\text{Forecast to go} = \text{Original unit rate} \times \text{Quantity to go}$$

a. This is a conservative approach if the current productivity rate is less than, or better than, the original unit rate.

b. This is an optimistic approach if the current productivity rate is more than the original rate, but the project manager assumes a learning curve or tool-up period is behind them and the productivity will improve.

3. Forecasting the current trending productivity rate times the quantity to-go. This assumes actual conditions are more accurate than the original estimate. This is only appropriate when a substantial amount of work has been complete and is not as applicable for early progress especially if a learning curve is to be considered.

Forecast cost (or hours) to go = Current productivity rate × Quantity to go

The same approach and options work for scheduling. For example, Figure 16.2 for the DFHW work package assumed Mountain Construction's foreman would average six door openings per day. At any point in time, based upon an as-built productivity rate, to-go productivity could be forecast and a final or revised schedule completion could be determined.

16.6 EARNED VALUE AS A PAY REQUEST TOOL

The concept of payments made on an earned value basis was originated by the United States Department of Defense in the 1960s and adopted to construction in the late 1970s. To utilize EV in the pay request process requires a very sophisticated and construction-experienced client. This process can benefit a contractor that is on a stipulated sum contract and ahead of schedule or under budget. The lump sum contractor can also be hurt by this process if they are over budget or behind schedule. Guaranteed maximum price (GMP) projects with a pre-established schedule of values operate very similar to a lump sum project if paid on an EV basis. This is assuming, although, that they are not subject to monthly audits of actual expenditures and corresponding adjustments to invoices.

Contractors with an open-book contract that are paid on a cost-plus fee or time and material basis are usually not subjected to earned value for pay requests. If they were being paid on an EV process, and they were over-spending, they would be penalized; conversely they would be rewarded if they were under-spending as the owner would need to pay them what they had earned which would be in excess of what was spent.

If a contractor on a cost-plus percentage fee project is paid on a purely cost-plus basis and not earned value, and they are under-running their estimate, their fee would also be potentially reduced as the fee in this case is based on a percentage of construction cost. Unfortunately, this results in a disincentive for contractors to beat the estimate. Conversely if they were over-budget, their fee would be increased in a pure cost-plus contract. This is why many cost-plus contracts include a high-side GMP protection for the project owner and "fix" the fee at the time the estimate is developed. These contracts are also known as cost-plus fixed fee.

16.7 SUMMARY

The use of earned value analysis requires a thorough understanding of the interplay of work package planning with estimating and scheduling, along with accurate reporting of costs expended and quantities installed. With only two curves, the budget and the actual curves, whether they be man-hours or dollars, it is not possible to measure true performance on either cost or schedule. This would not be possible until project completion. The third curve allows the team to answer the interim questions: Is the work package (or project) on budget, under budget, or over budget, and by how much, and is it on schedule, behind schedule, or ahead of schedule, and by how much? The whole project direct labor curve introduced in Chapter 10 was a rough way to monitor cost, but the work packages (Chapter 15) developed by foreman, which focus on a distinct assembly, usually organized by craft, such as carpenters versus electricians, is a much more accurate measurement. Now the third earned value curve is even more accurate – but it requires more involvement from the jobsite management team than the previous two.

The use of indices is a more technical EV analysis where formulas can be input into the computer and only percentage completion and cost or hours spent need to be added. Reporting only quantities is also an option. The use of indices without plotting the curve would also require additional input from the jobsite management team.

Forecasting final costs and schedule are an important element of project management. The most difficult element of forecasting is projection of the to-go costs or time. These forecasts can be subjective and project managers can take overly optimistic or pessimistic approaches – but a realistic approach is the best for reporting to upper management. Earned value methods can assist with monthly cost and fee forecast reports.

Very few clients use earned value as a method of payment, but actually invoicing lump sum projects with a preestablished schedule of values is very similar to EV. Using EV for pay requests is usually associated with experienced, sophisticated clients, such as government agencies for lump sum work, but not typically on privately financed negotiated cost-plus fee projects.

16.8 REVIEW QUESTIONS

1. What is an earned value analysis used for?
2. How is payment for a GC on an EV basis different than a lump sum basis with a schedule of values which was submitted on bid day and included in the contract as an exhibit?

3. If a GC was on a cost-plus project, without a GMP, how would payment on an EV basis affect them if they were (A) under-running their estimate, or (B) over-running their estimate?

4. If your project had estimated the foreman would spend 100 hours by day 4 of a six-day project, but she had only spent 80 hours, is your team over budget or under budget and ahead of schedule or behind schedule and by how much?

5. When would a contractor not want to use an EV curve for payment basis?

6. Which of the nine options in performance matrix Table 16.1 would be considered by a contractor to be completely neutral?

7. What are the calculated CPI, CV, SPI, and SV for DFHW work package Figure 16.2 as of day 8 and how does that compare to the results discussed in the narrative earlier?

8. What is the FCV and FSV for Figure 16.2 at completion of this work package?

16.9 EXERCISES

1. (A) Prepare an additional earned value curve starting with the work package example presented in Chapter 15, Figure 15.2. Status the work package as of day 8 based upon the "earned" information included in this chapter. (B) Complete the EV curve assuming all work is complete at the end of day 14.

2. Draw a separate EV curve where the foreman is over budget and behind schedule and another where the team is on budget and ahead of schedule. Prepare a narrative explaining why these situations might be occurring.

3. Refer back to the earned value curve in Figure 16.1. If this work package proceeds at the same trend as experienced as of day 8, when will it complete and what will be the final hours over or under budget?

4. In what option of matrix Table 16.1 would the two points plotted in Figure 16.3 be? What if the ACWP and BCWP points were in the opposite locations?

5. Have you worked on a project that required EV as either a cost and schedule control tool or pay request tool? Did it require a substantial amount of field management support?

6. Assume a contractor had 112 punch windows to install, and estimated an original productivity rate of four carpenters installing two windows per hour. Solve the following forecast questions:

   A. What was the original estimated productivity rate, duration and total man-hours for this work package?

   B. If after day 1, 10 windows had been installed, when will the work package be complete? What is the trending productivity rate? Provide an optimistic, pessimistic, and realistic solution. What will be the final productivity rates for each?

C. Answer the same questions in Exercise 6.B, except at day 3, 40 windows had been installed.

D. Answer the same questions in Exercise 6.B, except at day 5, 80 windows had been installed.

E. Extend the current trend from scenario 6.D to project completion. Utilize the appropriate wage rate from Chapter 10. How much money did this work package make or lose?

F. Explain what might have happened in this work package combining status points B, C, and D. You may need to draw an EV curve to help.

Subcontract Management

17.1 INTRODUCTION

General contractors (GCs) typically use subcontractors to execute most of the construction tasks involved in a project, often 80–90% of the scope of work. A GC that subcontracts 100% of the work is known as a pure construction manager (CM) as was introduced in Chapter 2. All, or 100%, subcontracted scopes and no direct GC work, may also be an owner requirement on a negotiated commercial project and is also typically the case with residential builders, especially speculative home tract contractors.

Subcontractors, often referred to as specialty contractors, therefore, are important members of the general contractor's project delivery team and have a significant impact on the jobsite team's success or failure. The relationship of the subcontractors to the other members of the project delivery team has been discussed throughout this book. Since subcontractors have such a great impact on the overall quality, cost, safety, and schedule success for a project, they must be selected carefully and managed efficiently. General contractor project managers (PMs) and superintendents find it advantageous to develop and nurture positive, enduring relationships with reliable subcontractors. Project managers and superintendents must treat subcontractors fairly to ensure the subcontractors remain financially viable as business enterprises. This will ensure that their subcontractors will be successful on this current project and be available for future projects.

Subcontractors provide labor and equipment for installation of materials at the jobsite. *Suppliers* fabricate and provide materials for installation at the jobsite, but the installation is by others, whether that be the GC or subcontractors. Subcontractors may also fabricate and supply their own materials. Subcontractors receive a subcontract agreement from the GC whereas suppliers receive a purchase order (PO). Most of the concepts in this chapter apply to both subcontractors and suppliers and the discussion will typically default to subcontractors for simplicity.

General contractors use subcontractors to both *reduce risk* and to provide access to specialized skilled craftworkers and equipment. One of the major risks in contracting is accurately forecasting the amount of jobsite labor required to complete a project. By subcontracting significant segments of work, the GC can transfer much of the risk to subcontractors. When the GC's estimator or PM asks a subcontractor for a price to perform a specific scope of work, the subcontractor bears the risk of properly estimating the labor, material, and equipment costs.

Subcontracting is not without risk. The GC's PM and superintendent give up some control when working with subcontractors. The scope and terms of the subcontract agreement define the responsibilities of each subcontractor. If some aspect of the work is inadvertently omitted, the GC is still responsible for ensuring the prime contract requirements are achieved. Specialty contractors are required to perform only those tasks that are specifically stated in their subcontract documents. Consistent quality control may be more difficult with subcontractors, particularly the quality of workmanship. Owners expect to receive a quality project and hold the GC accountable for the quality of all work whether performed by its own crews or by subcontractors.

Subcontractor bankruptcy is another risky aspect of subcontracting, which can be minimized by good prequalification procedures and timely payment for subcontract work. Scheduling subcontractor work often is more difficult than scheduling the general contractor's crews, because the subcontractor's tradesmen may be pulled off the site or committed to other projects. Safety procedures and practices among subcontractors may not be as effective as those used by the general contractor presenting an additional challenge to the GC. Because direct supervision of subcontractor craftsmen is the responsibility of the subcontracting firm's management, it can sometimes be frustrating for the GC's project superintendent to implement these CM control systems.

This chapter will discuss subcontractor selection, acquisition, and management. General contractors must remember that poor subcontractor performance will reflect negatively on their professional reputations and their ability to secure future projects. Once the subcontractors have been selected, subcontracts are executed documenting the scopes of work and the terms and conditions of the agreement. Subcontractor and supplier management is an integral part of construction management. While the GC's project superintendent manages the field performance of the subcontractors and suppliers, including schedule adherence, the project manager manages all subcontract documentation and communication and both field supervisors are responsible for ensuring the subcontractors are treated fairly.

17.2 SUBCONTRACTED SCOPES OF WORK

If the owner does not impose any subcontracting requirements on the general contractor, the GC may choose to self-perform scopes or award the work to specialty contractors for a variety of reasons, some of them are listed here.

- Subcontractor has access to specialized labor and equipment.
- Subcontractor can perform the work at a lower cost.
- Subcontractor can perform the work faster.
- Subcontractor assumes the risk with respect to quality control and potential rework.
- The GC can improve its cash flow by employing subcontractors that are subject to retention held from pay requests and are not paid until the owner pays the GC.

During development of the initial work breakdown structure, the project manager identifies the work items that are to be self-performed by the general contractor and those that are to be subcontracted. He or she develops a subcontract plan identifying which work items to include in each subcontract. This plan becomes the basis for creating the specific scope of work for each agreement.

The subcontract scope of work must specifically state what is included and what is excluded. A well-defined scope of work is essential to ensure prospective specialty contractors understand what is expected of them. Poorly defined scopes of work lead to conflicts and often result in cost escalation, time delays, and litigation. The subcontract agreement must state clearly the exact scope of work to be performed and either include or make reference to all relevant drawings and specifications. This ensures there will be a clear understanding of the subcontractor's responsibilities.

The specialty contractor's scope may include only construction services, or it may include both design and construction services. Construction services may include labor and equipment only or may include labor, materials, and equipment. Action words should be used in writing scopes of work to minimize misunderstandings. Words such as "design," "provide," and "install" should be used to define clearly the subcontractor's responsibilities, compared to "furnish," which is also applicable to material supplier purchase orders.

17.3 SUBCONTRACT DOCUMENTS

The subcontract agreement documents the understanding between the general contractor and each specialty contractor used on the project. It usually is based on the requirements and specifications contained in the general contract with the project owner. Once the scope of work has been established, the specific terms and conditions are defined for each subcontract agreement. These terms and conditions establish the operating procedures the GC's PM intends to use to manage each subcontract or PO. Topics that should be addressed include: Applicable drawings and specifications, commencement and progress of work, inspections, change order procedures, payment procedures, and many others.

Standard subcontract documents have been developed by the American Institute of Architects and ConsensusDocs. General contractors should choose one of these documents, because they are used widely in the industry and understood by specialty contractors. General contractors that choose not to use standard copyrighted documents should consult with legal counsel when crafting subcontract language. Creating subcontract agreements and POs from scratch is very risky and not recommended. General contractors will input language into these agreements that is beneficial to themselves and not necessarily beneficial to their new team members as shown in the next example. Subcontractors and suppliers should take the same precautions with contract review as has been suggested when a GC enters into a prime agreement with the project owner.

Example 17.1

The construction management company on this mega public-works project was experienced at managing and defending against subcontractor claims for schedule extensions. They inserted language in each subcontract agreement that indicated the project schedule would undergo literally 100 revisions in the three-year construction duration and each of these revisions was automatically incorporated into each subcontract and purchase order agreement. The revisions would routinely be posted to an electronic drop box and if the subcontractors did not respond within five days of a posting, they had accepted the revision at no change in cost. If you were a subcontractor or material supplier would you think this was fair?

17.4 SUBCONTRACTOR PREQUALIFICATION

To ensure quality work, estimators and project managers should prequalify specialty contractors before asking them to submit a price for the work. Historically, a heavy emphasis has been placed on price alone and not enough on schedule, quality, and safety records. General contractors should select subcontractors based on their ability to provide the greatest overall value to the project. Best-value subcontractor and supplier selection should be based on a standard set of criteria, including: Competency of field supervision; ability to meet schedule requirements; availability of specialized equipment and labor; safety record; and others.

One technique for gathering much of this information is to require prospective subcontractors to respond to a detailed questionnaire. Personal interviews with the GC's PM and superintendent are often required when selecting the set of prequalified subcontractors invited to submit proposals. Owners and designers on negotiated projects should also be invited to participate in reviewing the prospective subcontractors' qualifications. General contractors should use only quality subcontractors as members of their potential project delivery teams.

Early Release of Subcontractor and Supplier Bid Packages

One of the many services the general contractor brings to the preconstruction team is the ability to solicit pricing from subcontractors and suppliers who provide materials with inherent long-lead material delivery durations. General contractors rely on positive long-term relations with these companies to receive this valuable early input. Some of the subcontractors or suppliers the GC estimator and scheduler may engage early include: Mechanical, electrical, elevator, curtain wall, structural steel, design-build shoring, and others.

Ideally the general contractor will not issue subcontracts or purchase orders until it has also received a contract from the client. But sometimes these long lead materials and equipment must be ordered before that can happen. Some vendors will accept letters of intent that may offer financial reimbursement for early submittals and shop drawings and a promise of a forthcoming construction contract, if the project proceeds as planned. If a financial commitment has to be made to a subcontractor or supplier, then a similar commitment will need to be made from the project owner to the contractor and incorporated into the preconstruction fee and contract. The firms engaged with early bid packages will also input firm submittal and material delivery information into the contractor's detailed schedule. This is a form of collaborative scheduling as discussed further on in this chapter.

17.5 SUBCONTRACTOR SELECTION

Selecting quality specialty contractors for each subcontract scope is essential for project success. Quality specialty contractors have good safety and schedule records, experienced craftspeople, good equipment, and adequate financial capability to complete the project to the desired standards without experiencing financial problems. Some GCs select subcontractors simply on price. This often leads to problems on the project with quality control or timely execution.

Many general contractor superintendents feel quality and safety procedures and practices of subcontractors are not as effective as those used by the general contractor's own workforce. All of these construction management control aspects are even more crucial when subcontractors also subcontract portions of their work to third-tier subcontractors and suppliers, as shown in the Figure 17.1 organization chart. It is imperative that the GC selects quality "best-value" subcontractors, at every tier, if the GC's project manager and superintendent are to produce a quality project, on time, safely, and within budget.

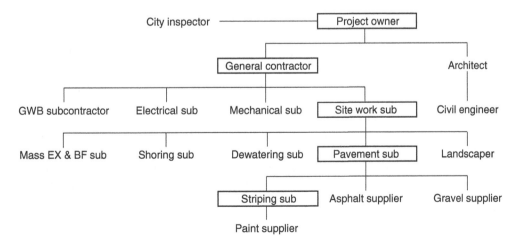

Figure 17.1 Third-tier subcontractor organization chart

The recommended strategy is to select the top-ranked five or six specialty contractors identified during prequalification and invite them either to submit a proposal or a bid for the scope of work. Requests for proposal (RFPs) typically are used for design-build or cost-plus scopes of work. Invitations to bid or requests for quotation (RFQs) are used for lump sum or unit price scopes of work.

Once the proposals or bids have been received, they are evaluated to select the best specialty contractor for each scope. After each specialty contractor is selected, the price is entered on the subcontract, and the agreement is signed both by the GC and the specialty contractor. The detailed contract schedule should be reinforced in each PO and subcontract agreement. The procedures used are similar to those discussed in Chapter 2 for executing the contract between the owner and the general contractor. A sample ConsensusDocs 751 subcontract agreement for the fire station overhead and apparatus bay doors is available on this book's companion website. All these steps are considered part of a diligent subcontractor and supplier buyout process.

17.6 TEAM BUILDING

Successful construction projects require collaboration among many parties representing the project owner, the designer, general contractor, subcontractors, suppliers, and regulatory agencies. The GC's project superintendent does not build a project by him or herself, they need a team of qualified experienced professionals. Assembling a reliable group of subcontractors and direct craft foremen requires collaboration among the GC's management team, including the PM and superintendent.

Successful superintendents have select foremen who follow them from project to project. Foremen also have a select group of skilled craftsmen they call on to make them successful. An understanding of what trades do which work is a requirement for a superintendent, and bringing all of these diverse men and women and companies together into a project team is an essential leadership skill that takes time to acquire. The GC's project superintendent must partner with foremen, subcontractors, the owner, design team, city, and community groups in order for the project to be successful. Today's general superintendent must build a team around him or herself of reliable direct craft foremen, subcontractors, suppliers, and craftsmen from a variety of trades.

Because specialty contractors make up such a substantial share of the workforce on a typical commercial project, the only way for a general contractor to achieve success is to also make the subcontractors successful. Choosing only best-value subcontractors is a big step towards that success. The GC's superintendent should be present at buyout and preconstruction meetings to understand commitments made with respect to manpower and schedule, and should have access to each subcontract agreement at the project site. The subcontract agreement is an important construction management tool that solidifies all of the planning work that contributed to a vendor's selection.

17.7   SUBCONTRACTOR MANAGEMENT

Once the subcontracts are awarded and construction begins, the general contractor's project manager and the superintendent work together to schedule and coordinate the subcontractor's work to ensure the project is completed on time, within budget, and in conformance with contract requirements. Since most of the construction work is performed by subcontractors, efficient management of their work is critical to the jobsite team's ability to control costs and complete the project on time. Coverage requirements should be stated in the subcontract. Before any subcontractor is allowed to mobilize onto a project site their certificates of insurance must be in place. Additional premobilization requirements should include the submittal of a site-specific safety plan, execution of the subcontract agreement, attendance at the subcontractor preconstruction meeting, and others.

As discussed above, the first challenge is to mold the subcontractors into a cohesive project delivery team. This requires an understanding of their concerns and proper work sequencing to ensure their success. The GC's superintendent should require each subcontractor to weekly submit short-interval schedules and daily reports of its activities. This provides an ongoing record of each subcontractor's progress and any obstacles encountered in performing the work. The requirement to submit short-interval schedules and daily reports must be included in the terms and conditions of each subcontract agreement.

Collaborative Scheduling

Top-down scheduling is frequently encountered on many projects, especially public projects based on lump sum or unit price contracting that rely on low bidding for prime contractors and subcontractors. This approach to construction scheduling is not a good match with best-value procurement. Top-down scheduling means the client has dictated an end date and the general contractor has to meet it, at all costs, or likely suffer liquidated damages. The GC may use the same process with their subcontractors. In some cases, the GC will have the schedule produced by an in-house staff scheduler or scheduling consultant, without necessarily involving the field management team. Often these schedules have little detail and are difficult to use on the jobsite as a construction management tool. Once the GC's superintendent receives a top-down schedule with only the contractual end date noted, he or she may in turn pass this same process down to their subcontractors and foremen and put the burden on them to achieve schedule success.

Does the general contractor schedule the subcontractors or do the subcontractors schedule themselves? The same principle of commitment and collaboration applies to *subcontractor developed schedules* as it does to superintendent developed estimates and schedules discussed earlier in this book. If the subcontractor develops their own schedule, or at a minimum input to the GC's schedule, and if it fits within the overall plan, they have made a commitment to achieve the schedule. One method to develop collaborative schedules includes pull planning as discussed previously with lean construction planning. Whereas, if the GC dictates in a top-down fashion to the subcontractors when and how long they will be on the job, the subcontractors may accomplish the task, but if they don't finish on time, they will always have the excuse that they did not get the opportunity to provide input.

The ConsensusDocs 751, Standard Short Form Agreement between the Constructor and a Subcontractor, has several articles that connect with this collaborative scheduling approach. This family of documents was created from a consortium of contractors, including GCs, CMs, and subcontractors. The language is not one-way, as is reflected in excerpts from the following articles.

> *Article 1: The subcontractor shall provide work in accordance with the progress schedule to be prepared by constructor <u>after consultation with the subcontractor</u>, and as it may change from time to time.*
> *Article 3: The general contractor's progress schedule will be <u>included</u> as a contract exhibit to this subcontract agreement.*
> *Article 7.2: The schedule development and maintenance is a <u>collaborative effort</u> between the constructor and the subcontractor as <u>they both work together</u> with progress and updates . . . (Underscores added for emphasis).*

A valuable schedule tool as discussed previously in this book is a *specialty schedule*. The mini-schedule or fragnet can be focused on just one subcontractor, such as an earthwork subcontractor or the curtain wall subcontractor. A specialty schedule can be customized or simply printed through the use of scheduling software filtering, which is a special sort with just one subcontractor's work displayed. The shortcoming of the "sort" method is that it does not necessarily include the interplay with the GC or other subcontractors before, during, and after the focused subcontractor's efforts. A specialty schedule which focuses on one subcontractor and includes all other related activities, is a more comprehensive tool, but requires a little more creativity on the part of the scheduler. A specialty schedule for the fire station roofing subcontractor is included on the companion website.

Fast-track scheduling was a popular term utilized in the 1980s. Today, most construction projects are scheduled and built on a fast-track basis, which means that many activities are overlapping and occur concurrently in lieu of sequentially. Very few schedules are completely linear in that all the preceding activities are 100% complete before the subsequent activity starts. Fast-track design and construction allows expedited start dates and turns the project over to the client faster, which reduces construction loan interest and improves developer performance. Overlapping activities allows the GC and its subcontractors more flexibility in balancing manpower; they have more activities to choose from. However, fast-tracking a schedule reduces leeway, so it is important that all team members are aligned around the schedule.

Similar to the estimate, the schedule must be developed in *sufficient detail* to allow project progression to be adequately measured, monitored, and corrected if necessary. Too much detail and the schedule will take on a life of its own and becomes too burdensome to monitor. The 80-20 rule should be applied to schedules the same way it is applied to estimates: 80% of the work or time is included in 20% of the activities. The project team should focus on those critical 20%. Providing just the right amount of detail on a schedule is a way to reach out to foremen and subcontractors and improves their work planning efforts and development of three-week look-ahead schedules.

In example 17.1, the construction management company included language in the subcontractor's agreement that would imply the subcontractor would have to work *overtime* or *off-shift work* in case there was schedule compression. Schedule compression is often a source of delay claims, as introduced in the next chapter. Some subcontracted scopes may be performed more efficient off-shift or on weekends due to additional unobstructed access – e.g. painting, acoustical ceilings, fire proofing, floor covering, and others. The best solution for off-shift work is to choose a specific task or work area or subcontractor, such as a painter, as shown in the next example.

The best solution for a general contractor to work with its subcontractors, now often referred to as "trade partners," is through team building and collaborative scheduling. A new schedule should be reviewed with the critical subcontractors before it is published. *New or revised schedules* should be walked through by the GC's super-intendent and scheduler with all of the subcontractors upon issue and not just mailed with a contract modification requiring incorporation and execution without input.

Subcontractor Controls

All subcontractors must be aware of their *safety responsibilities* including those respect-ing safety for other subcontractors working on the site. The GC's superintendent must ensure that the jobsite is ready for a subcontractor before scheduling them to start work. Requiring subcontractors to arrive on a site that is not ready causes a hardship for them and can result in lost time and potential increased project cost. The super-intendent also is responsible for resolving any conflicts between subcontractors. Even though the work is being performed by a subcontractor, the GC is still responsible.

Example 17.2

Working multiple shifts is not always a way to recover schedule time as there are typically inefficiencies to overcome. This industrial project included "high bay" structural steel and catwalks that required painting after all of the mechanical and electrical work was installed. This steel was 80 feet above the slab-on-grade. The painting subcontractor requested the GC allow them to work on a second shift, without additional compensation, so that the painters had the entire project to themselves. The GC had to staff the second shift with another superintendent for insurance and safety reasons, but he kept himself busy updating drawings with as-built information.

The superintendent must ensure subcontracted work conforms to the *quality requirements* specified in the contract. Subcontractors must understand workmanship requirements before being allowed to proceed with the work. The GC's superintendent should conduct *preconstruction meetings* with subcontractors before allowing them to start work on the project. The owner and designer should be invited to participate in all subcontractor preconstruction meetings, especially on negotiated projects. Often mock-ups, which are stand-alone samples of completed work, are required for exterior and interior finishes. This allows the superintendent and the designer to evaluate the work and establish a standard of workmanship. Mockups are just one form of *submittal* that are required by suppliers and subcontractors, the others being shop drawings, material cut sheets or product data, and samples. The importance of submittal management as part of schedule control is discussed in detail in Chapter 14. Project *site cleaning* policies also should be discussed at the preconstruction conference. Generally, all subcontractors should be required to routinely clean their work areas when they have completed their scopes of work.

Situations often arise when the scope of work needs to be modified. All such modifications should be documented as *subcontract change orders* to identify clearly the changes in scope and the impact on the subcontract price. Sometimes changes affect multiple subcontractors. In these cases, all affected subcontract agreements must be changed appropriately. Prior to negotiating the cost of a change order, the project manager sends a change order request to the affected subcontractors.

Subcontractors submit *requests for payment* as their work progresses, at the end of each phase of their work, or according to the payment schedule provided in the contract. Payments should be made within the timeline established in the subcontract agreement. Most subcontracts contain a provision that the subcontractor will be paid for the work performed once the owner pays the general contractor. Such contracts place significant financial burden on subcontractors when owners fail to make timely progress payments to general contractors. Subcontractors should be paid timely to avoid causing them cash flow problems and to ensure their financial health. Before making final payment, the subcontractor should be required to submit a final lien release.

Before a subcontractor demobilizes from the project the general contractor's project superintendent must make sure that all their field work is complete, which includes punch list corrections, testing and inspection, and permit sign-offs. There are many other *close-out activities* that the GC's project manager will engage the subcontractor for, including as-built drawings, operation and maintenance manuals, final change orders and pay requests, and unconditional lien releases. These are all topics of project management books like *Management of Construction Projects* and the interested reader should look to a source, such as that for additional PM responsibilities associated with subcontractor management and controls, including schedule control.

Supplier Management

Supplier management is similar to subcontract management except the supplier enters into a purchase order agreement in lieu of a subcontract agreement and does not provide any direct jobsite field labor. Material suppliers also have labor involvement but that is associated with off-site fabrication and delivery efforts. As relates to scheduling, the GC's field management team will track all material deliveries and provide continual, often weekly, expediting oversight. This effort typically begins as soon as the PO is executed. Each supplier is assigned to one of the GC's field management team, such as the project engineer, project manager, superintendent, and even an administrative assistant. These people will routinely call or email each supplier to assure that the materials will be delivered as scheduled. Each of these delivery dates were backed into from the detailed contract schedule such that the materials arrive just in time to support installation – not too early and not too late – which is a lean construction planning principal. Subcontractor and supplier deliveries will be planned and managed in an expediting log or schedule similar to Table 17.1.

17.8 SUMMARY

Subcontractors are essential members of the general contractor's team. Typically, they perform most of the work on a construction project. During the initial work breakdown for a project, the GC determines which work items are to be subcontracted and which of those work items to include in each subcontract. Based on this subcontracting plan, the project manager crafts a specific scope of work for each subcontract agreement. Once the scope of work has been developed, the GC prepares the terms and conditions of the subcontract or selects a standard subcontract format.

Because subcontractors are so critical to project success, general contractors should prequalify them before asking for price quotations. Subcontractors should be selected based on their ability to provide the greatest overall value as members of the project delivery team.

Once the subcontracts are awarded and construction begins, the general contractor's project manager and the superintendent work together to schedule and coordinate the subcontractors' work to ensure the project is completed on time, within budget, and in conformance with contract requirements. The success of the project is dependent on a viable schedule that provides adequate notice to all subcontractors regarding the scheduled start times for their phases of work. General contractors that work with their subcontractors to develop a collaborative detailed schedule often complete their projects on time. Subcontract management includes ensuring quality performance of work, responding to requests for information, issuing change orders when needed, and promptly paying subcontractors for accepted work.

Table 17.1 Material expediting schedule

No.	Description	Supplier	Committed submittal	Actual submittal	Fabricate/ delivery duration	Scheduled delivery date	Actual Delivery Date	Responsible person
1	Import sample	Earth Enterprises	5/29/22	6/1/22	NA	6/18/22	6/12/22	Pam Sampson
2	Pea gravel	Earth Enterprises	5/29/22	5/28//22	NA	6/18/22	6/13/22	Sampson
3	Silt fence	Tri-Cities Lumber	5/29/22	5/31/22	NA	6/18/22	6/10/22	Sampson
4	Perforated pipe	Tri-Cities Lumber	6/5/22	6/2/22	2 days	6/18/22	6/16/22	Sampson
5	Footing rebar	Steel Fabrication	6/5/22	6/7/22	2 weeks	7/10/22		Charles Kent
6	Embed steel	Steel Fabrication	6/5/22	6/7/22	2 weeks	7/10/22	6/19/22	Kent
7	Concrete mix	University Redi-Mix	6/5/22	6/1/22	NA	7/10/22		Ralph Henry
8	SOG wire mesh	Steel Fabrication	6/15/22	6/14/22	1 week	8/5/22	7/4/22	Kent
9	Asphalt mix	ATB Asphalt		7/5/22	NA	2/17/23		Henry
10	Steel trusses	Steel Fabrication	6/10/22	7/5/22	8 weeks	9/25/22		Kent

17.9   REVIEW QUESTIONS

1. Why do general contractors use subcontractors rather than performing all the work on a project with their own labor forces?

2. What risks do general contractors incur by using subcontractors?

3. Why might a project manager require performance and payment bonds from a subcontractor?

4. Why do subcontractors need reasonable notice regarding the scheduled start times for their phases of work? Why does the superintendent not determine the exact start times for each subcontractor at the beginning of the project?

5. What are mock-ups, and how are they used for project quality control?

6. Why is a subcontractor's ability to finance his or her cash flow requirements on a project of concern to the GC?

17.10   EXERCISES

1. Why is a termination or suspension provision included in most subcontracts?

2. Why is subcontractor bid shopping considered unethical behavior?

3. How are subcontractors' requests for payment processed on a typical project?

4. What is the difference between a RFQ and a RFP? What type of subcontract is awarded using a RFQ? What type is awarded using a RFP?

5. Write a clear scope of work for a subcontract for the site work scope for the Fire Station 83 case study project.

6. What are five criteria that you suggest be used to prequalify roofing contractors for the case study project?

7. What basis do you suggest the project manager use to short-list electrical subcontractors bidding on the case study project?

Chapter 18

Schedule Impacts

18.1 INTRODUCTION

This last chapter in this part on schedule controls, which is also the last chapter in the book, introduces many advanced topics that impact planning and scheduling. Many of these topics are also connected with other materials discussed previously. Most of them would warrant their own chapters in books on construction management (CM), but here the reader is provided with just an overview of these advanced topics. All of the topics in this chapter impact the way contractors plan, schedule, and control construction projects. The sections in this chapter include:

- Time value of money: The adage that "time is money" applies to the construction industry, maybe more so than in other industries. It will cost a contractor more to build tomorrow than it will today, so achieving the schedule is paramount to achieving financial success.
- Time and cost trade-offs: Contractors will calculate the trade-offs of building faster or slower and the cost-benefit analysis of each. Both approaches can cost more or save more money and warrant careful analysis before implementation.
- As-built schedules: Recording when activities started, how long they actually took, and when they finished enhances the scheduler's ability to prepare better schedules in the future. As-built schedules are also essential for claim preparation and defense.
- Claims: If a contractor's schedule is impacted by others, the workforce loses productivity and costs are increased and the contractor looks to recover from the project owner.
- Legal impacts: There are many terms and concepts in construction contracts involved with claim preparation and defense that are connected to scheduling.
- Risk management: Construction projects are often run as autonomous business centers by a project manager (PM) and superintendent. These leaders must be able to understand and mitigate risks in order for their company to be successful.
- PERT and other scheduling methods: This book has focused on construction scheduling tools most often utilized by jobsite teams, but a few other advanced scheduling methods also warrant a brief introduction.

18.2 TIME VALUE OF MONEY

The concept of time value of money (TVM) is basically defined as the benefit of having a stipulated sum of money today is greater than the benefit of having the same amount of money in the future. TVM is a connection between the importance of a positive cash flow for contractors that is enhanced by fair and timely pay requests and negotiating and receiving payment for change orders. Money costs money, and if a project owner makes a late payment or negotiations are slow on change orders, if the project is delayed, or if there is an unfavorable amount of retention held, this will affect a contractor's cash flow. Time value of money is a popular topic in the study of engineering economics.

Time value of money is an advanced construction management and scheduling concept and is not necessarily project-based, but related to the roles of chief executive officer and chief financial officer (CFO). This topic is still very relevant to jobsite managers; it is important that the jobsite team understands that "money costs money." If a contractor needs to borrow money from the bank to pay their monthly labor and material invoices, they will be operating with a negative cash flow, or in the red. It would be unusual to find a general contractor (GC) including the cost of a construction loan in their estimate. This can be validated by looking at the detailed case study jobsite general conditions estimate and the expanded live general conditions template, both included on this book's companion website. There is not a line item for a construction loan in either of those documents. The contractor has no intention of financing the project for its client. If a PM operates in the red and has a negative cash flow, he or she will quickly be receiving a phone call or visit from the CFO. A construction application of TVM is more relevant to larger construction projects with longer durations than smaller and faster projects.

These financial terms are often used synonymously but have slight variations in definition, particularly as related to time value of money, including:

- Price: The amount that is or was asked;
- Cost: The amount that was paid, or given up;
- Value: The worth an object has to someone, not necessarily the price or the cost; and
- Worth: The value that someone else is willing to pay.

One important premise of time value of money is the concept of "equivalence." A study of TVM makes analysis of former, current, and future investments as financially equivalent as possible. TVM does not factor nonquantitative elements, such as wants and wishes and personal preferences. In order to compare sums of money in two different time frames an understanding of present value (PV), future value (FV), interest, inflation, and other value considerations is necessary. A brief introduction of these TVM terms follows. An understanding of TVM helps companies and individuals when analyzing schedule impacts because "time costs money."

Present value is today's worth of an investment, e.g. an investment of $100 made 20 years ago is presently worth $1,000. Present value is also today's equivalent, say $500 of a future value of $1,000 seven years from now, assuming the investor could realize an established interest rate or return on today's $500 investment. *Future value* is the forecasted worth today's investment has in the future depending on anticipated inflation rate. The FV or today's $1,000, 30 years in the future, may be $2,500. Future value and PV are inverses of one another.

Interest is the cost of money, or the amount one party (the bank) will pay another (a bank depositor) to use their money. One is the borrower and the other the lender. The use of money is not free. Compounding interest allows the investor to make interest on top of interest. *Inflation* measures the historical changes of values. If bread cost $2.00 per loaf in one year, and $2.50 in another year, that 50-cent increase reflects a 25% price increase, or 25% inflation.

The concept of time value of money is simple; $100 yesterday was worth more than it is today, and today's $100 will be worth more than it will be tomorrow. There are many formulas related to present value, future value, annuities, interest, and inflation. Most of the combinations have already been figured out for the TVM novice and are available on websites. In addition to the internet, the interested reader can look to a resource like *Cost Accounting and Financial Management for Construction Project Managers* for additional tables, calculations, and equivalence discussion for all of these concepts.

18.3   TIME AND COST TRADE-OFFS

Contractors must weigh trade-offs between time and money. There is an optimum construction duration in which the project is the most cost efficient. If the project takes longer than planned to build it costs more money in labor inefficiencies and extended jobsite overhead, including equipment rental and others. If the project is expedited or has a shorter duration it also costs more money due to excessive overtime, productivity impacts, poorer quality, and potentially an unsafe construction project. The trade-off of time to cost efficiency is reflected in Figure 18.1. Regardless of the amount of money that can be thrown at a project, there is some finite minimum duration for the building process.

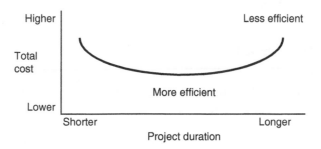

Figure 18.1   Efficiency versus time trade-off

Costs on a construction project that are typically time-dependent are primarily jobsite general conditions costs, also known as indirect costs. If a project lasts longer, the contractor has more time to do the work, which should theoretically improve efficiency, but it also increases indirect labor and material or jobsite overhead costs. The longer a project lasts, the more is spent on jobsite general conditions, typically those specific items that are variable costs (time-dependent) versus fixed. Some of these costs include:

- Superintendent and project manager and other field management wages;
- Rental on office trailers;
- Rental on construction equipment, such as cranes, forklifts, and pickup trucks; and
- Temporary site utilities including power, water, phones, and heat.

Conversely projects which have a shortened duration saves jobsite overhead costs but the compressed timeframe often reduces labor efficiency due to congested work areas, increased labor cost due to overtime premiums, and likely jeopardizes jobsite quality control and safety control. These impacts were discussed in Chapter 10.

Working a second shift can also negatively impact labor efficiency. Additional shifts also increase onsite supervision costs as a second foreman or superintendent must supervise nightshift work. Some contractors will work their second shift only 7.5 hours, but pay them for eight, often working through their lunch break. Another approach is to increase the wages of the second shift by 10%. If it is a union project, there are different rules for off-shift premiums based upon the craft. Some of the other potential negative impacts of a second shift include:

- Redo work,
- Duplicate work,
- Communication issues,
- Poor lighting,
- Safety problems,
- Security problems, and others.

18.4 AS-BUILT SCHEDULES

An as-built schedule, as is an as-built estimate, is an important historical tool that experienced contractors utilize to successfully procure future construction projects. The time to start development of the as-built schedule is when the schedule control process started, with development of the detailed schedule that is hung on the conference room wall in the jobsite trailer. During the course of construction, material delivery dates and labor hours are accurately recorded, along with measurement of actual quantities installed, to develop as-built labor productivity rates. The as-built schedule is definitely part of the construction close-out process, but the sooner it is prepared, the more accurate the results will be. These durations should be shared by the project team with the staff scheduler for incorporation into the contractor's database. In this manner, the last phase of schedule control, the as-built phase, is actually the first phase of the next project, and the schedule control cycle depicted in Figure 8.1 repeats itself.

Schedule control involves monitoring the progress of each scheduled activity and selecting appropriate mitigation measures to overcome the effects of any schedule delays. As-built schedules are prepared to allow PMs to develop historical productivity factors for use on future projects. An example portion of the case study's as-built schedule, which has been marked-up in hand by the superintendent, is included as Figure 18.2. Schedules should be maintained in a manner similar to the way record

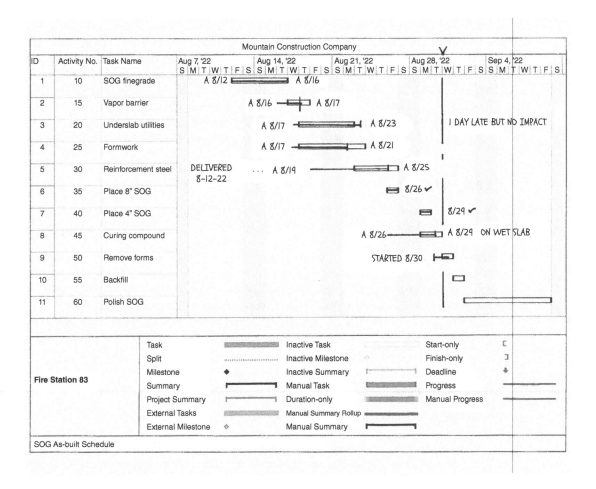

Figure 18.2 As-built schedule

drawings and as-built estimates are maintained. They should be contemporaneously marked up and commented upon to reflect when materials were delivered, activities started, activities finished, rain days, crew sizes, and whatever else the project manager and superintendent feel is appropriate. Heavy civil unit price projects also may record actual material quantities installed.

Just as an as-built estimate is important for estimating future work, an as-built schedule is helpful in scheduling future projects. Similar to estimating, the greatest risk in scheduling is the determination of direct craft workforce productivity. The project manager and/or home-office staff scheduler should develop a personal set of productivity factors based on actual prior experience and historical data. These factors will help future jobsite project teams establish realistic activity durations when scheduling their next projects.

The marked-up detailed jobsite meeting room schedule is a permanent record. It includes backup as to why delays occur. The GC's superintendent's daily diary, as well as subcontractors' diaries, also record planned versus actual manpower, deliveries, and impacts to the plan. Some contractors continuously revise their detailed schedule with actual start and completion dates. In this way the last revision of the schedule also serves as their as-built schedule. Often the main use of an as-built schedule is backup for a claim. As-built schedules are also prepared by a claims consultant or investigator – someone who was not out on the project. The as-built schedule, similar to the daily job diary, becomes a legal tool in the case of a dispute. The use of as-built schedules and daily job diaries as construction tools is introduced in Chapter 14.

18.5 CLAIMS

Sometimes issues occur on construction projects that cannot be resolved among project participants. Such issues from a contractor's perspective typically involve requests for additional money and/or time for work performed beyond that required by the construction contract. The project manager first submits a change order proposal (COP) for a proposed contract adjustment. If the owner and the PM agree, a change order is negotiated and executed, adjusting the contract duration and/or price. A claim will therefore not be filed. If the owner does not agree with the project manager's request, the result may become a contract claim. This section on claims is further divided into the following subsections: Sources of claims, claim prevention, claim preparation, claim defense, and claim resolution.

Sources of Claims

Construction claims typically result from one of the following causes:

- Constructive acceleration, or compression, where the contractor is required to perform a task or tasks at a faster rate or with additional manpower than planned.
- Constructive changes where the scope of work was modified by the owner.
- Cumulative impact of numerous requests for information and change orders.

- Unresolved change orders.
- Defective or deficient contract documents.
- Delay caused by the owner or the owner's agents, including inspectors and the design team.
- Site conditions that differ from those described in the contract documents.
- Unresponsive contract administration.
- Force majeure.
- Weather impacts and others.

An industry partner summarized construction delays with the following: "Costs incurred as scheduled do not have any impact. Costs incurred later than scheduled have a potential impact and may result in a claim."

Any of these issues may affect the jobsite team's ability to complete the project within budget and in the prescribed duration. The major issues in claims and disputes are:

- What are the issues?
- Who is responsible?
- What does the contract dictate?
- What are the cost and time impacts?

Time impacts are determined by assessing the impact of the disputed work on the construction schedule. Many of the above sources could impact the construction schedule. Delay claims stem from an unexpected event; the contractor could not have anticipated it and therefore expects to be excused from any associated time and cost

Example 18.1

On a recent project, a sizable subcontractor refused to provide any schedules of any value during the entire two years they were on the project. Remarkably, two months after completion, the subcontractor submitted a claim for extra costs incurred due to schedule delays. An exhibit to that claim was an incredibly detailed schedule, in color, computer generated, with floats, deliveries, and manpower restraints all indicated that had not been shared while the job was in progress. The reason this happened was because the schedule did not exist until after the fact; it was developed by a professional claims consultant. This schedule was not a construction management tool. If it had been available early in the project, maybe the subcontractor would have managed their work properly and communicated early with the GC, and a claim would not have been required.

impacts. Cost impacts are determined by evaluating the actual costs for any additional work plus the overhead costs incurred by the contractor for being on the project any extra time. Unfortunately, many schedules today are used largely for claim preparation, as shown in Example 18.1.

Force majeure contract clauses excuse the contractor from performance as these could not have been anticipated and there would be little the contractor could do to mitigate their impacts. Force majeure literally means acts of god; some examples include:

- Hurricanes,
- Tsunamis,
- 100-year storm conditions, such as excessive snow, ice, hail, and freezing weather,
- Volcano eruption,
- Tornadoes, and
- Earthquakes.

This does not include a few extra rain or snow days. A few of these should have been planned for and considered reasonable, as shown in Figure 18.3, which is an excerpt from the special conditions to the contract from another public works lump sum bid project.

General contractors may be impacted on their ability to finish a project on time and within budget for almost limitless possibilities. Even though a project owner may not recognize many of these as a legitimate delay claim, it does not restrict the GC from attempting to recover for its losses. Some additional schedule impacts a GC may experience include:

- Delayed issue of city permits or changes in the design as a result of the permits.
- Delayed delivery of specified materials beyond the GC's control, including owner-provided materials or equipment.
- Poor performance of subcontractors, including the owner's subcontractors.
- Poor quality or safety performance of any of the team members.
- Delayed submission of submittals by suppliers or delayed return by the design team.
- Change of personnel from any of the team members.
- Financial strain or bankruptcy on the part of any of the team members, including subcontractors.

The ConsensusDocs 500 negotiated contract, Article 6.3.1 includes a long list of examples of delay causes beyond the control of the construction manager; most of

Special conditions of the contract

Claims for delays due to abnormal weather

Weather delays for normal weather conditions will not be considered. Non-compensable delays for adverse weather may be considered if it can be shown that the weather was unusually severe and outside of the conditions established in a table similar to below for the project's location.

Month:	Rain days > 0.1"	Cold days < 25 deg. F.
Jan	16	3
Feb	14	2
Mar	14	1
Apr	13	0
May	10	0
June	9	0
July	8	0
Aug	7	0
Sept	8	0
Oct	13	1
Nov	16	2
Dec	16	3

Figure 18.3 Abnormal weather specification

them are typical and noted previously, such as weather, but it also adds terrorism and epidemics to the list. This book is being written during the 2020/2021 Covid-19 pandemic, so there will be no surprise that pandemics are added as an excusable delay to standard contract language in the future.

Claim Prevention

At times, it seems attorneys are always lurking around construction projects looking for claim opportunities. The PM's and superintendent's role in keeping accurate schedule documentation throughout the construction process plays a crucial role in both claim preparation and claim defense.

The partnering and collaborative team-building techniques discussed in Chapter 3 and others have been adopted by many in the construction industry in an attempt to reduce the number of claims and disputes. The frequent meetings and the issue escalation system are used to resolve issues at the project level in a timely manner. Some readily-apparent methods to prevent or mitigate claims are included in the next list. Projects that are claim-free often have these characteristics.

- Experienced project owner,
- Good design team and good design documents,
- Minimal design changes after construction starts,
- Selection of only a competent GC and best-value subcontractors,
- Experienced GC project manager and superintendent,
- Fast monthly payment procedures,
- Expeditious change order resolutions,
- And, most important, good communication procedures.

Claim Preparation

Procedures for processing claims typically are prescribed in the contract. Articles 9.4 and 13 of the ConsensusDocs 500 contract outline the procedures used for a CM at-risk delivery method. The normal procedure is for the GC's main office to formally submit the request for additional compensation and/or time to the owner or the architect along with documentation supporting their position.

Requiring contractors and their craftsmen to work for extended periods of overtime, for whatever the cause, is often a reason for a post-project claim. Everyone agrees that overtime can take its toll and reduce productivity as discussed earlier, but it is difficult to measure that effect and calculate and agree upon the cost impacts. Even management can be affected by excessive overtime requirements, as shown in the next example.

A contractor's formal claim may have many of the same elements as does a change order proposal except it would likely be more extensive and for more money and/

Example 18.2

Everyone pretty much agrees that working extended periods of overtime has a detrimental effect on direct craft productivity, but it is the magnitude of that effect that will be debated. Not only does excessive overtime affect the quality of work, it also has an impact on the health of the workforce. In this specific case, the scheduler also worked seven 12-hour days per week for a solid year. He would get home late in the evening and would receive a phone call early the next morning to get back to work on a mega public works project. Although he enjoyed the extra pay, it affected his health and his family life to the point that he left that lucrative job at the conclusion of the year.

or more time. Some of the major cost areas the contractor would look to the project owner for recovery include:

- The actual cost of added work;
- Loss of productivity for contract work;
- Extended jobsite general conditions cost;
- Home office overhead costs (see Eichleay further on);
- Loss of fee potential; and
- Similar cost areas for subcontractors.

The Fire Station 83's detailed estimate and jobsite general conditions are located on the companion website. Note that there is not a detailed home office general conditions estimate included in the project estimate as home office general conditions are considered part of any contractor's fee and are not cost-reimbursable as explained in most contracts. But this does not prohibit a contractor from attempting to claim these costs.

It is important to back up the baseline schedule – it often finds its way into claim discussions. Contractors compare an as-planned versus as-built schedule and expect the project owner to compensate for 100% of the difference. Comparing the difference between the baseline schedule and the as-built schedule is often the backup for a request for additional time. Other backup for claims include the superintendent's daily diaries, weekly owner-architect-contractor meeting notes, monthly schedule updates or revisions, monthly executive reports, and date-stamped photographs of completed work.

Some schedulers focus on collecting actual data solely for the purpose of preparing a post project cost and/or time claim. Having enough versus too much detail in a schedule is a double-edged sword. Without the detail it is difficult for a contractor to prove the exact impact of an interruption – although many will acknowledge that there may have been some impact. Without enough detail in a schedule, the project owner could say "That time or activity should have already been included or was inferred in another longer activity." Conversely a schedule which is overly-detailed can be held over the contractor's head in case they also miss an internal deadline. Overly detailed schedules take on a life of their own and may require a full-time on-site scheduler to maintain them. If the project superintendent is the only team member responsible for schedule maintenance, he or she becomes tied to the schedule and is not managing field productivity, as was shown in the previous Example 12.2.

While conducting research for this book this author came upon the term *delay log*. Another scheduler recommended contractors keep such a list with the obvious intention of submitting a post-project time and/or cost delay claim. This is not a typical document or practice, especially in private work. All contracts require a contractor to bring any delay claim issue to the attention of the project owner when it occurs and not accumulate several into one global claim. Doing so does not allow the owner and/or its agents to respond in a proactive manner to mitigate whatever may be the cause of these delays. Planning for a claim and maintaining a delay log seems counter-productive to the team approach to construction.

Claim Defense

The owner or the architect formally responds to the contractor's request for a change order agreeing, agreeing in part, or rejecting the contractor's proposal. If the contractor does not agree with the response, the claim becomes a contract dispute and the dispute resolution technique discussed in the contract is used to settle it.

Many project owners will utilize contractual technicalities to defend against claims, even if the claim had some merit. How and when a contractor serves 'notice' to an owner (and a subcontractor serving notice to a GC) is often contended more than the cause of the extra cost and/or time. The AIA A201 general conditions, Article 15.1.2 indicates that "notice must be given within 21 days of occurrence."

Concurrent delays involve two or more delays, one potentially excusable and another non-excusable. It may be difficult to separate the time associated with each, but is necessary if the contractor expects any recovery. Any request for extra time from a contractor must separate contractor-caused delays from owner-caused delays; often these two are blended, sometimes accidently and sometimes intentionally. This was the case with earlier Example 13.2. If any portion of the claim is potentially the contractor's responsibility, the project owner may use this as a justification for rejection in part or all.

Excusable delays allow the contractor to collect extra time, but depending on other contract language, as discussed further on, may not increase their contract dollar amount for jobsite or home office general conditions. A non-excusable delay is deemed to have either been caused by the contractor, and therefore potentially negligent, or should have been anticipated and the contractor is not entitled to either time or monetary compensation. A "no claim for delay" contract clause attempts to protect the project owner from extended overhead costs associated with delays. A compensable delay awards the contractor extra time and money but must be associated with an excusable versus non-excusable delay.

Contractors are expected to mitigate disruptions to the schedule; often contract language requires this. If it is proven a contractor did not attempt to work around disruptions, caused by them or another party, or amplified a disruption, they would likely lose their opportunity to claim for impact. An accelerated schedule, or recovery schedule, are ways contractors may catch up or simply continue on after impacts, whether internally or externally caused.

A fallacy exists on the part of scheduling consultants, owners, designers, attorneys and others that contractors have additional cost and time built into their bids and schedules to accommodate changed conditions. Contractors do not include a time contingency that they can simply remove to make up for added scope or discovery of unknown conditions. This is similar to contractors not including a cost contingency in their estimate when preparing a competitive bid. If they did so, and their competition did not – which they don't – they would not become low bidder.

Claim Resolution

It is to the advantage of both the contractor and the owner to resolve all disputes quickly. The farther up-the-ladder above the project level a dispute reaches, the more likely it is to become adversarial, time consuming, and expensive. The most economical way to resolve a claim is to (A) first prevent it from occurring, and (B) negotiate resolution at the jobsite level.

The prescribed claim resolution methods specified in most construction contracts include: mediation, arbitration, and litigation. Usually either mediation or arbitration is indicated but not both. The Fire Station 83 contract stipulates the following order of dispute resolution between the GC and project owner:

- The architect makes the first determination;
- The dispute is subjected to mediation if the parties do not agree with the architect;
- If mediation does not resolve the dispute, the issue is elevated to litigation.

Recent standard AIA and ConsensusDocs contracts have added another choice, which is a third-party neutral or the assembly of a Dispute Resolution Board (DRB) which participates throughout the course of construction such that the DRB remains informed if or when a dispute materializes.

During a dispute, the settling parties, either the court or arbitrator, must consider whether the original as-planned schedule was correct to begin with in regards to durations and restraints. If a claim gets to the litigation stage, then judges, juries, and attorneys who are not necessarily versed in construction processes and terminology will eventually make the decisions for contractors and project owners. In such cases scheduling consultants will draft after-the-fact critical path method (CPM) schedules, as discussed in the earlier example. The court system typically prefers detailed, and sometimes complicated, CPMs over simple bar charts that do not necessarily include relationships. Detailed schedules and records also have more weight in claims and court than summary or generic schedules.

Both parties bring in outside scheduling consultants and expert witnesses in a legal dispute over construction time, including this author. It is not surprising that they all have different opinions, as each expert witness is paid to represent a different party.

18.6 LEGAL IMPACTS

As indicated, claims differ from change orders in that negotiated and executed change orders resolve, and incorporate into the contract, new or changed conditions. Claims often involve more money and time impacts than individual COPs. Some of the industry trends and concepts that impact the size and resolution of claims follow. Many of these are contract considerations the GC should evaluate before proceeding with a bid or proposal, as was discussed in Chapter 6.

- Waiving rights to claim: There can be clauses in the base contract, lien releases, pay requests, and executed change orders where GCs and, in turn, subcontractors are required to waive their rights to future claims. Different jurisdictions and courts may either uphold these clauses or deem them unlawful.

- No claim for delay: This is a contract clause that allows the owner or designer to delay a project, due either to additional scope or delayed decisions, but the contractors are not allowed to recover additional jobsite or home office overhead costs, even though they may be given time extensions on the contract schedule. The AIA A201 general conditions, Article 15.1.6 includes this standard verbiage: "If a delay is caused by any party other than the contractor and/or its subcontractors, the contractor is entitled only to an increase in time but not a change in the contract sum."

- Eichleay formula: Eichleay is a complicated and controversial element of some claims. This formula calculates the home office overhead and company fee potential that is applicable to an individual project. If a project is delayed at no fault of the contractor, then in addition to recovering jobsite general conditions costs, home office costs and lost fee are added to the claim as well. Many standard contracts disallow the use of Eichleay.

- Case law: Construction disputes are civil and not criminal lawsuits. Attorneys and the judge will rely on prior court cases to base their position, and ultimately the decision on a construction claim. If a court found in favor of the contractor on unknown site conditions on one project, that finding may be applied to a future project as well.

- Notice requirements: Notice was discussed in Chapter 6 regarding contract considerations and it is applicable to claims as well. If a project owner denies a contractor's COP, then, as dictated by contract, the contractor has a set amount of time, say 30 days, to notify the owner that they disagree with the disposition of the change order and intend to file a claim. The claim itself will also need to be formally served on the owner within a given timeframe, and/or referred to mediation or arbitration, whichever is called for in the contract. Essentially the contractor cannot wait until the end of the project, or later, to bring up old issues.

Following are additional terms and concepts that affect both development and resolution of many construction claims:

- Quantum vs. merit: The merit or validity of a claim should be analyzed before the quantum or amount or cost of the claim is weighed;
- Excusable vs. non-excusable: If the contractor caused the problem, they are deemed non-excusable and cannot claim for additional time and/or money;
- Compensable vs. noncompensable: A compensable claim is one that the contractor deems eligible for financial remedy;
- Culpable or culpability: The degree one of the parties may be partially or fully responsible for delay or cost impacts;
- Avoidable vs. unavoidable: Could the contractor have mitigated or avoided the situation or did they proceed into it with the intention of claiming additional cost and/or time?
- Schedule compression: Often more work is being requested in the same time period;
- Schedule acceleration: Shortening the duration of the project either because of delays or the addition of scope.

18.7 RISK MANAGEMENT

Risks are plentiful on any construction project. Many of these are external risks and beyond the jobsite team's control. Some others are internal risks and the GC's home office expects the jobsite team to manage them. The home office executive team first evaluates risks when they decide whether or not to bid or propose on a project. Construction project risks include:

- Variability of labor, material suppliers, and subcontractors;
- Subcontractor performance and financial solvency;
- Weather and other conditions beyond the field management's control;
- Quality of design;
- Conditions imposed by the city or permitting agency;
- Each project is unique;
- Contract terms;
- Subcontractor claims, default, and potential bankruptcy;
- Schedule adherence and avoidance of liquidated damages;
- Cost control;

- Safety control;
- Quality control; and
- Equipment operation.

Construction risk management process needs to be a team effort and involves the following steps:

- Identify the risk,
- Transfer the risk,
- Eliminate the risk,
- Accept the risk,
- Mitigate risk, and
- Monitor progress and measure success.

The jobsite team has the greatest ability to control, manage, or influence construction cost and time with direct labor, which is also the riskiest cost and schedule category. Direct labor is one of the most difficult costs to estimate and the most difficult to control, but also provides the contractor with an opportunity to improve its fee if the team beats the estimate. The GC's PM and superintendent may also identify other areas or systems of the project that might be risky and warrant additional attention during estimate and schedule development and controls. Risk management begins with the proposed contract language review before a bid or proposal is tendered, as was discussed in Chapter 6. Some of these risk areas might include:

- Existence of contaminated or hazardous materials,
- Congested areas where multiple crews will need to work at the same time,
- Areas with difficult access,
- Portions of the work completely dependent upon subcontractor success,
- Unfamiliar scope areas or areas outside of the contractor's normal expertise,
- Close proximity to neighbors,
- Environmentally sensitive areas,
- Performance of project owner supplied materials and/or subcontractors;
- Existing utility conflicts, and
- Lack of design detail.

18.8 PERT AND OTHER ADVANCED SCHEDULING METHODS

Program evaluation and review technique (PERT) was also originated by the U.S. military, but is rarely if ever used in construction. Even most scheduling authors and experts pretty much agree that PERT schedules do not have a place in construction. PERT schedules track three dates, the best or *most optimistic date* or duration, the worst or *most pessimistic date* or duration, and the *most-likely date* or duration. The mean or most-likely duration is calculated utilizing statistics, standard deviations, probability calculations, areas under curves, certainly versus uncertainly ratios, and others. In reality, contractors rely on what seems to be their 'best guess' but is likely an educated guess based on past experience, material quantities, confirmed delivery dates, and input from expert builders, such as superintendents and subcontractors. The PERT scheduling concept calculates the probability of achieving a duration or date and assumes that contractors have done this prior. This may be appropriate with industrial manufacturing or speculative home developments, but not commercial, industrial, or custom home construction projects. PERT is generally not a good communication or construction tool.

Monte Carlo simulations are similar to PERT schedules. In this case the scheduler, or statistician, simulates the process and eliminates the unlikely options, leaving the most likely, which therefore makes up the schedule. Similar to PERT, the scheduler makes many assumptions and assigns a probability percentage to each option. Monte Carlo is more likely found with owner program evaluations rather than a GC schedule on a construction site.

Velocity diagrams are essentially a graph that charts quantities of work scheduled and installed on a vertical *y* axis and time on the horizontal *x* axis. For some, these are also considered schedules and may appear on heavy civil projects, such as highways, tunnels, utilities, and piling – all with repetitive operations.

Linear scheduling utilizes concepts such as modeling and velocity diagrams and likely is also an appropriate tool for production schedules utilized in manufacturing industries, such as car manufacturers. The linear scheduling method (LSM) graphs times versus locations, distances, and/or quantities. These schedules may also be known as time change diagrams. The use of linear scheduling is not common in commercial construction. It may have an application in repetitive projects, such as civil projects with long stretches of fairly straight roads or pipelines as well as speculative home projects where the same five models are being built over and over. It may also have some use with construction claims but the focus of this book has been schedules as tools to build a project, not one to prepare a lawsuit. While conducting research for this book I found one author who stated "linear scheduling is obviously understood and useful in construction." After my nearly 50 years in a variety of roles from residential carpenter to cost and schedule engineer to project manager, company executive and owner, professor and author, owner's representative and expert witness, I personally have never seen a linear schedule used on a commercial, industrial, or residential construction jobsite to help build the project. But my good friend who focuses on heavy civil public projects, such as utilities and light rail, finds LSM a valuable tool – especially from an owner's perspective. He contributed LSM Figure 18.4.

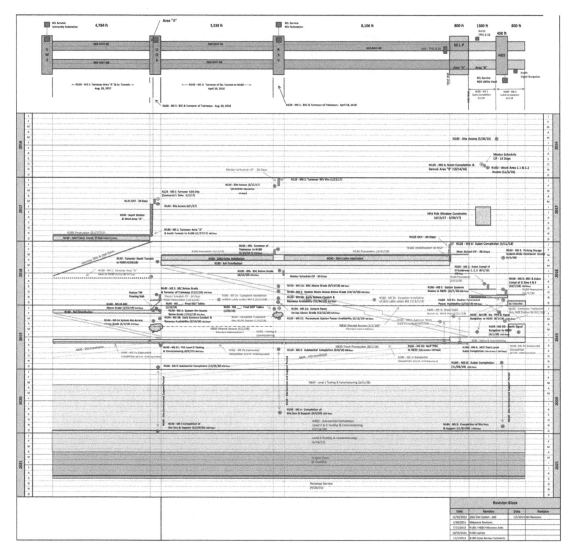

Figure 18.4 Linear schedule example

18.9 SUMMARY

This is the last chapter of Part IV, project controls. This chapter covered a variety of impacts the schedule has on construction management and CM has on the schedule. Time value of money simply means "money costs money." Today's dollars will not be as effective in the future. It will cost more money to build a project a year from now than it does this year; therefore, schedule adherence is paramount for the contractor

There is an optimum duration for each activity and each project. Contractors must be able to compare cost and time trade-offs when evaluating schedule alternatives. Expediting and/or delaying construction may both result in increased cost exposure.

An as-built schedule records when each activity started, how long it took to do the work, and when it completed. Recording actual events like material deliveries and inspections should also be included on the contractor's as-built schedule. This tool can be useful for future schedule developments or claim backup. Claims are often the result of unapproved change orders. Claims are typically for more time and cost impact than change orders. Claims exist on many projects, especially those bid lump-sum, and can be unfortunate occurrences for all of the team members. Claims can be expensive to prepare, defend against, and resolve. Claim avoidance is the most economical solution and the best option in order to retain team relationships. But a claim may be necessary and is a tool that every contractor should have in its toolbox.

If a claim materializes the contract will determine which dispute resolution method the parties are to follow. Claims and lawsuits involve an additional set of terminology – much of which is included in the contract agreement. Contractors should be well versed on legal aspects with respect to claim preparation, defense, and resolution before they proceed with a bid or a proposal.

Some methods of risk management by the home office include purchasing insurance, contingency funds, higher fees, joint venturing, bonding, choosing best value subcontractors, and utilizing the right PM and superintendent for the right project. There are a variety of other scheduling methods, including PERT and LSM. These options are likely better-suited for a professional scheduler or claims consultant and are not typically seen out on the jobsite. This book has focused on schedules as tools to help build the project.

The other chapters in this final part of the book focused on a variety of schedule control methods available for contractors, including:

- Schedule control involves strategies of when it is necessary to revise a schedule or prepare a new recovery schedule versus updating the current schedule. Short-interval schedules are one of the most powerful tools for the last planners in the field.

- Scheduling tools include incorporation of submittals into the detailed schedule.

- Jobsite controls cover a variety of construction management topics including cost control, quality control, safety control, and document control.

- Earned value management is another method of cost and schedule control, which introduces a third curve to the typical cost and schedule control work package. Earned value measures work in place compared to cost incurred to accomplish that work.

- Subcontractors account for 80–90% of the workforce on most commercial construction projects. It is imperative that the GC's project manager and superintendent effectively manage their subcontractors to achieve project success.

The other three parts of the book led up to schedule control. An introduction to construction management, including contracts and organizations was included in Part I. Part II covered a variety of scheduling planning topics including preconstruction services, lean construction planning, and contract considerations every estimator and scheduler should consider before submitting a bid or proposal. Development of a realistic logic diagram, along with associated restraints, is the result of good planning.

The next group of topics in Part III converted the plan into a schedule communication tool. This involved adding durations and timelines to the logic diagram, along with calculating the project duration and determination of the critical path. Those activities not on the critical path have float and there are a variety of methods schedulers and contractors choose to display and manage potential float. Resources play a critical part in schedule development and management. Some of the resources a contractor focuses on include the availability of qualified field labor, construction equipment, and cash needed to finance the project between the time work is incurred and the time payment is received from the project owner. Schedule software tools have allowed schedulers to prepare schedules faster, perform calculations, incorporate schedule updates or revisions, and produce a variety of reports.

I hope you enjoyed the connection this book has made to schedules as useful tools on the construction jobsite. Maybe one day you, too, will become a 'scheduler.'

18.10 REVIEW QUESTIONS

1. Why might it be less expensive to build a project (A) one month faster and/or (B) one month longer?

2. The cost of a cubic yard of redi-mix concrete is more expensive: today, five years ago, or five years from now?

3. What is included in an as-built schedule?

4. What is the difference between a change order and a claim?

5. How are claims resolved?

6. Which four of the following legal terms would a contractor need to have in its favor to stand a good chance of collecting on a claim: Excusable, non-excusable, compensable, noncompensable, culpable, nonculpable, avoidable, and/or nonavoidable?

7. Spell out at least four of these abbreviations: DRB, TVM, CPM, PERT, PV, CFO, FV, COP, LSM, and/or AIA.

18.11 EXERCISES

1. (A) Today's value of $1,500 was worth $____, 25 years prior assuming a 5% compounding interest rate? (B) What would the value of today's $1,500 be 25 years

into the future assuming a 3% compounding inflation rate? Utilize TVM tables from the internet.

2. What additional items from Mountain Construction's fire station general conditions estimate located on the companion website are time-dependent?

3. Have you ever worked on a swing shift or a graveyard shift? Were you paid more? How was your performance compared to working a dayshift?

4. Can you think of any additional risks that can be added to this chapter's list provided in the risk management section? Of those you added, separate them into internal or external risk categories.

5. Assume you are the project manager for a construction company with an average annual volume of $750 million, which spends 1.25% on home office overhead and routinely realizes 2.5% in pre-tax pure profits. Your own project is scheduled to last two years and has a contract value of $100 million, 6% of which are for jobsite general conditions. Midway during construction your client has delayed your project for one year at no fault of your own. Prepare two claims: (A) Jobsite overhead costs: Can you recover 100%? What would you do with your people and equipment during this one year shut-down? (B) Utilize the Eichleay Formula to prepare a claim for impacts to home office overhead and lost fee potential.

6. Draw a schedule utilizing the Fire Station 83 photographs included in the book. Assume today's date as day one. Utilize the relationships and durations from the detailed schedule included on the companion website with the progress accomplished in the photographs. Make whatever assumptions as needed.

7. Combine Webster's definitions of planning, scheduling, and control and create one connected definition. Does that fit with construction? How might you modify it so that it fits with our conversation? If you come up with a good solution, please send it to holmcon@aol.com and I will incorporate in the next edition of this book, with credit to you!

Glossary

3D: three-dimensional drawing where a third dimension, depth or height, has been added to the traditional two-dimensional drawings; a popular tool of CAD and BIM.

4D: the addition of a fourth dimension, time, to three-dimensional drawings.

5D: the addition of a fifth dimension, cost, to three- and four-dimensional drawings.

Active quality control or active quality management program: preparing plans and taking actions to prevent mistakes before they happen rather than repair them after they have occurred.

Active safety control: a process that anticipates and prevents safety problems rather than just responding to and correcting issues after an accident; see also *project-specific safety plan*.

Activity: smaller defined element of a path or project with start and end dates and a duration, consumes resources; also known as a task.

Activity-based costing: the process of applying home office and also jobsite indirect costs to departments and projects and direct construction activities if possible.

Activity-based resourcing: the process of applying resources such as labor and equipment to direct work activities.

Activity duration: the estimated length of time required to complete an activity.

Addenda: additions to or changes in bid documents issued prior to bid and contract award; plural: addendum.

Administrative restraints: nondirect construction activities to be added to a detailed schedule, including submittals, permits, material deliveries, and others.

Agency construction management delivery method: a delivery method in which the client has three contracts: one with the architect, one with the general contractor, and one with the construction manager. The construction manager acts as the client's agent but has no contractual authority over the architect or the general contractor.

Agreement: a document that sets forth the provisions, responsibilities, and the obligations of parties to a contract. Standard forms of agreement are available from professional organizations such as the American Institute of Architects, ConsensusDocs, and others.

AIA A201: general conditions for the conventional AIA family of contract documents that set forth the rights, responsibilities, and relationships of the owner, contractor, and architect.

American Institute of Architects: a national association that promotes the practice of architecture and publishes many standard contract forms used in the construction industry.

Application for payment: see *payment request*.

Apprentice: a beginner tradesman, typically in their first one to seven years of the trade and still in training. He or she possibly also attends classes outside of the job in a technical college. Advances to journeyman.

Arrow-diagramming method: a technique that uses arrows to depict activities and nodes to depict events or dates; also known as activity on the arrow diagram.

As-built drawings: contractor-corrected construction drawings depicting actual dimensions, elevations, and conditions of in-place constructed work.

As-built estimate: an assessment in which actual costs incurred are applied to the quantities installed to develop actual unit prices and productivity rates.

As-built schedule: a marked-up, detailed schedule depicting actual start and completion dates, durations, deliveries, and restraint activities.

As-planned schedule: could be considered the *contract schedule* or *baseline schedule* but is sometimes developed to support a claim to compare with the *as-built schedule*.

Assemblies analysis: determining the number of man-hours estimated per unit of work, such as man-hours per cubic yard of concrete, and comparing the result with similar data from other projects.

Assemblies cost estimate: a *semi-detailed cost estimate* that requires quantity takeoff of bulk in-place systems, such as foundations, that have been sufficiently designed, and applies a unit price for all work associated with that system, including labor, material, and subcontractors. Systems that are not yet adequately designed may be estimated by plugs or allowances. May also be associated with a GMP estimate.

Assignable: appoint or place responsibility, such as assign plumbing fixture installation to the plumbing subcontractor.

Associated General Contractors of America: a national trade association primarily made up of construction firms and construction industry professionals.

Back charge: a general contractor charge against a subcontractor for work the general contractor performed on behalf of the subcontractor.

Backwards pass: adding together the length of schedule activities, starting with the final activity and working backwards toward the first activity, all the while calculating the late start and late finish of each activity.

Balancing: a form of ductwork and plumbing testing and adjustments.

Bar chart schedule: a time-dependent schedule system without nodes that may or may not include restraint lines.

Base hourly wage: that portion of a worker's gross hourly pay that does not represent deductions for fringe benefits or payroll taxes.

Baseline schedule: the original schedule, may also be the contract schedule, the schedule for which subsequent schedules will be compared.

Bid documents: drawings, specifications, and contract terms issued to general contractors to be used for developing a bid for the defined work. Bid documents are not always the same as the final construction documents.

Bid form: the form issued by a project owner on which contractors submit their bids.

Bid procurement process: selection of a contractor based upon a lump-sum bid; see also *procurement methods*.

Bid shopping: unethical general contractor activity of sharing subcontractor bid values with the subcontractor's competitors in order to drive down prices; also known as bid peddling.

Bluebeam: software facilitating construction teams to prepare quantity take-offs and develop sketches which accompany requests for information, and other uses.

Budget cost estimate: a preliminary estimate based on early design documents; see also *rough order of magnitude cost estimate*.

Budget options log: a spreadsheet used to monitor changes of materials or scope throughout the design process and the budget implications of those changes; also known as budget control log.

Building information models or modeling: computer design software involving multidiscipline three-dimensional overlays that can improve constructability and reduce change orders.

Built environment: facilities and physical infrastructures that add or change functions to the underlying natural, economic, and social environments.

Buyout: the process of awarding subcontracts and issuing purchase orders for materials and equipment.

Calendar schedule: action items or construction activities are placed on a traditional calendar format; popular as a *short-interval scheduling* tool.

Cash flow: the amount of revenue coming into a company from which the amount of expenses is subtracted. A positive cash flow represents the company is in the black and a negative cash flow (more expenses than revenue) shows that the company is operating in the red.

Cash flow curve: diagram charting cost versus time and reflecting the amount of money spent or expected to be spent; product of a *cost loaded schedule*; also known as a cash flow schedule.

Certificate of occupancy: a certificate or document issued by the city or municipality indicating that the completed project has been inspected and meets all code requirements.

Certificate of substantial completion: a certificate signed by the client, architect, and contractor indicating the date that substantial completion was achieved. Commonly used forms are AIA G704 and ConsensusDocs 814.

Chain: see *path* or *link*.

Change order: modifications to contract documents made after contract award that incorporate changes in scope and adjustments in contract price and/or time. Commonly used forms are AIA G701 and ConsensusDocs 202.

Change order proposal: a request for a change order submitted to the client by the contractor, or a proposed change sent to the contractor by the client requesting pricing data.

Change order proposal log: a listing of all change order proposals indicating dates of initiation, approval, and incorporation as final change orders.

Chief executive officer: the individual at the top of a company's organizational chart; may also be the president or *officer-in-charge*.

Chief operating officer: a senior company executive who is responsible for all construction operations; usually reports to the CEO; PMs and superintendents report to the COO possibly through middle-managers such as SPMs or general superintendents.

Chief scheduler: the lead scheduler or boss, assuming a company has more than one scheduler, typically a home office employee, may also be known as the *staff scheduler*.

Claim: an unresolved or postproject request for a *change order*.

Close-out: the process of finishing all construction and paperwork required to complete the project and close-out the contract; also includes project management close-out, financial close-out, and construction close-out.

Close-out log: a list of all close-out tasks that is used to manage project close-out.

Collaborative scheduling: inclusion by the GC of its subcontractors into development of a team schedule, different from a *top-down schedule*.

Commencement: day one or the beginning of a schedule or construction project.

Commissioning: a process of testing and assuring that all equipment and operating building systems are working properly, especially MEP systems .

Compensable delay: the contractor is reimbursed for costs incurred for a schedule delay beyond their control. Typical costs to be reimbursed are jobsite general conditions.

Completion: the last activity or day of a construction schedule or project.

Conceptual cost estimate: cost estimates developed using incomplete project documentation; also known as a *schematic cost estimate* or *budget cost estimate* or *rough-order-of-magnitude cost estimate*.

Conceptual design: initial design developed during the planning phase as part of the conceptualization of the project idea and evolved during the initial stage of design activities; may also be known as project concept .

Concrete schedule: spreadsheet scheduling or tracking delivery of concrete to the jobsite.

ConsensusDocs: a family of contract documents that has taken the place of the AGC contract documents.

Constraint: dates imposed on a contractor by a project owner such as a completion date or the city with road openings or shutdown dates or times; may also be known as *milestone* dates; should be shown in the *contract schedule*.

Constructability: the ability of a designed project to be buildable as designed.

Constructability analysis: an evaluation of preferred and alternative materials and construction methods; also a design quality control process.

Construction documents: a larger set of documents beyond the *contract documents* that includes a multitude of CM communication tools, including RFIs, COPs, meeting notes, daily job diary, and many others. May also be known as the conformed set of documents.

Construction drawings: the portion of the contract documents providing a graphical description of the project, including its geometrical information and often its materials .

Construction manager: when referred to an individual, this term is used to identify the contractor's leadership position on a project. When referred to firms, this term is used to identify companies that provide construction management services and serve a similar role to general contractors except they do not typically self-perform scope and, therefore, do not hire direct craft labor.

Construction manager agency delivery method: see *agency construction manager delivery method.*

Construction manager-at-risk delivery method: a delivery method in which the client has two contracts: one with the architect and one with the construction manager/general contractor. The general contractor usually is hired early in the design process to perform preconstruction services. Once the design is completed, the construction manager/general contractor constructs the project; also known as CM/GC delivery method.

Construction manager/general contractor delivery method: see *construction manager-at-risk delivery method.*

Construction schedule: see *contract schedule.*

Construction specifications: the portion of the construction documents that provide a textual description of the project, including any additional information on its materials, quality acceptance processes or other performance expectations. Construction specifications are usually organized according to some standardized classification system. For instance, the *Construction Specification Institute* (*CSI*) *MasterFormat* and *UniFormat* are two commonly-used classification systems for organizing data about construction requirements, products, and activities.

Construction Specifications Institute: the professional organization that developed the original 16-division MasterFormat that is used to organize the technical specifications; today's CSI includes 49 divisions.

Contract or contract agreement: a legally enforceable agreement between two parties.

Contract documents: the agreement, general conditions, special conditions, drawings, and specifications.

Contract schedule: schedules provided to the client at the beginning and throughout the project delivery as required by the contract special conditions. They are typically referred to in the contract agreement and are therefore contract documents. They are also typically exhibits within subcontract and purchase order agreements; also known as *formal schedules.*

Contract time: the period of time allotted in the contract documents for the contractor to achieve substantial completion; also known as project time.

Cost codes: codes established in the firm's accounting system that are used for recording and tracking direct and indirect costs.

Cost control cycle: a systematic financial expense control process engaged by a contractor starting with a construction estimate progressing through buyout, cost recording, process modifications, and culminating in an as-built estimate that feeds back into the company database and supports the next project estimate.

Cost estimating: the process of preparing the best educated anticipated final cost of a project, given the parameters available.

Cost-loaded schedule: a schedule or spreadsheet in which the value of each activity is distributed across the activity's duration, and monthly costs are summed to produce a cash flow curve.

Cost-plus contract: a contract in which the contractor is reimbursed for stipulated direct and indirect costs associated with the construction of a project and is paid a fee to cover profit and company overhead; also known as a time and materials contract.

Cost-plus contract with guaranteed maximum price: a cost-plus contract in which the contractor agrees to bear any construction costs that exceed the guaranteed maximum price unless the project scope of work is increased.

Cost-plus-fixed-fee contract: a cost-plus contract in which the contractor is guaranteed a fixed fee irrespective of the actual construction costs.

Cost-plus-percentage-fee contract: a cost-plus contract in which the contractor's fee is a percentage of the actual construction costs; also known as a time and materials contract.

Cost-reimbursable contract: a contract in which the contractor is reimbursed stipulated direct and indirect costs associated with the construction of a project. The contractor may or may not receive an additional fee to cover profit and company overhead.

Craftsmen: the nonmanagerial field labor force who construct the work, such as carpenters and electricians; also known as craftspeople or *tradesmen*.

Critical activity: any activity that occurs along the *critical path*; a delay in start or completion of a critical activity delays the critical path of the project and ultimately the final completion date.

Critical path: the sequence of activities on a construction schedule that determine the overall project duration. A delay in any one of the activities on the critical path results in a corresponding delay in the final project completion date.

Daily job diary: a daily report prepared by the superintendent that documents important daily events including weather, visitors, work activities, deliveries, and any problems; also known as daily journal or daily report.

Delivery or delivery method: the project owner chooses the contractual arrangement between the principle architect and the general contractor and how the design and the construction will be contractually 'delivered' to the project owner.

Demobilization: physically shutting down the construction project and removing the site camp and all construction equipment.

Dependencies: one schedule activity is dependent upon, or has a *relationship* with another activity, usually the *predecessor*; see also *successors*.

Depreciation: the loss of value of an asset such as a piece of construction equipment or building due to normal wear and tear that can be deducted from other profits to reduce income taxes; also equipment and/or real estate depreciation.

Design-bid-build delivery method: a delivery method in which the project owner hires a single contractor who constructs the project based on the design provided by the project owner; the design is substantially completed before a contractor is selected through a bid process.

Design-build delivery method: a delivery method in which the client hires a single construction company that both designs and constructs the project.

Design-build-operate maintain delivery method: a delivery method in which the contractor designs the project, constructs it, and maintains and operates it for a period of time, e.g. 20 years.

Design development documents: a design package advanced beyond the SD phase that includes outlining the specifications, e.g. architectural information, such as floor plans, sections, and elevations, as well as layouts of structural, mechanical, electrical, and plumbing systems. These drawings and specifications, when they are about 75–80% complete, may be the set of documents used when the building permit is applied for and/or a GMP estimate is prepared.

Design phases: three to five sometimes distinct and sometimes overlapping efforts to produce design documents with each phase becoming more accurate and more complete. Includes programming, conceptual, schematic, design development, and construction documents.

Detailed schedule: a schedule that is often hanging on the wall of a meeting room or the jobsite trailer and is marked up with comments and progress.

Differing site condition: some condition of the project site that is materially more adverse than as depicted in the contract documentation and could not be seen by a visit to the site. For example, encountering a buried water line where none was shown in the contract drawings.

Dispute: a contract claim between the owner and the general contractor that has not been resolved.

Dual occupancy: substantial completion is achieved for part or all of the project and the client has begun use, even though the construction team still has work activities in other parts of the project; also known as beneficial occupancy or joint occupancy.

Dummy: a false activity without duration or requiring resources that is used to show relationships and/or restraints between two other activities; typically associated with *arrow-diagramming scheduling method*.

Duration: the time it takes, in days or weeks or months, for an activity to occur.

Early finish: the earliest time at which a scheduling activity can finish.

Early start: the earliest time at which a scheduling activity can start.

Earned value: a technique for determining the estimated or budgeted value of the work completed to date and comparing it with the actual cost of the work completed. Used to determine the cost and schedule status of an activity or the entire project. Usually involves creation of an 'earned' or third curve, beyond estimated and actual curves; also known as earned value method, earned value approach, or earned value management.

Earned value indices: formulas and ratios to compare estimated cost and schedule to actual cost and schedule utilizing the third earned value curve.

Eichleay formula: a complicated method of potentially recovering home office overhead and lost fee potential usually associated with claims involving time extensions.

Eighty-twenty rule: on most projects, about 80% of the costs or schedule durations are included in 20% of the work items; also known as Pareto's Principle.

Environmental compliance: contractor and project owner's requirements to comply with city and state regulations, often requires an environmental plan.

Equipment schedule: a spreadsheet to track equipment use on the project including maintenance activity and potentially costs.

Estimate schedule: a management document used to plan and forecast the activities and durations associated with preparing the cost estimate. Not a construction schedule.

Estimator: a person charged with preparing a cost estimate; may also be the chief estimator or the *project manager.*

Event: as related to scheduling, this is the start date or time of an activity and also its finish or end date or time; the event does not consume time or resources as does an activity.

Excel: see *Microsoft Excel.*

Excusable delay: a time extension granted to a contractor by the project owner for events beyond the contractor's control.

Expediting: process of monitoring and actively ensuring vendor's delivery compliance with the purchase order requirements.

Expediting log: a spreadsheet used to track material delivery requirements and commitments, also expediting schedule.

External risk: issues contractors, project owners, and designers must consider into their plans and actions, these issues being beyond their direct control.

Fast-track construction or schedule: overlapping design and construction activities so that some are performed in parallel rather than in series; allows construction to begin while the design is being completed; also known as phased construction.

Fee: contractor's income after direct project costs and jobsite general conditions costs are subtracted from revenue. Generally includes and also known as a combination of home office overhead costs and profit.

Field engineer: similar to the *project engineer* except with less experience and responsibilities. May assist the superintendent with technical office functions.

Field question: see *request for information.*

Filter: the ability for scheduling software to remove or block activities outside of the requested sort, such that only one area of work or subcontracted scope might be printed or displayed from a larger schedule.

Final completion: the stage of construction when all work required by the contract has been completed.

Final inspection: the final review of a project by the owner and architect to determine whether final completion has been achieved.

Finish schedule: an Excel spreadsheet that lists every room within the building and includes columns for floor finishes, wall finishes, ceiling finishes, and others. The finish schedule is often accompanied with a key which describes symbols or abbreviations utilized in the schedule, not a time-related schedule.

Finish to finish: term that relates the completion of two activities, the completion of one activity is connected or related to the completion of another activity.

Finish to start: one scheduling activity has to be finished and is a predecessor to another, which cannot start until the first is completed.

Float: the flexibility available to schedule activities not on the critical path without delaying the overall completion of the project.

Flow of work: one activity or scope or subcontractor leads to the next without interruptions or work stoppages.

Forecast: monthly cost estimate and report prepared by the *project manager* for upper management that combines costs incurred with costs to go and results in a new estimated fee. Each monthly forecast should be more accurate than the last. Also known as fee forecast, cost and fee forecast, monthly forecast, and/or monthly project management forecast.

Foreman or foremen: direct supervisor of craft labor on a project.

Foreman work packages: see *work packages*.

Formal schedule: may be developed and provided to the project owner as required by contract or prepared and submitted with a proposal for a negotiated contract. This schedule may also be the *contract schedule*.

Forward pass: a calculation of the earliest start and earliest completion dates of all of the schedule activities in a network beginning with the first activity and ending with the last.

Fragnet: a partial view or sort from a larger schedule, potentially focused on one area of work, such as the second floor, or one subcontractor, such as the earthwork subcontractor.

Free float: the maximum amount of time a scheduling activity can finish late without impacting the start of the next successor activity; also known as activity float.

Front loading: a tactic used by a contractor to place an artificially high value on early activities in the *schedule of values* to improve cash flows.

Future value: forecasted worth an object has in the future depending on anticipated inflation rate; see also *present value* and *time value of money*.

Gantt chart: *bar chart* named after Henry Gantt in the early 1900s.

General conditions (1):   a part of the construction contract that contains a set of operating procedures that the owner typically uses on all projects. They describe the relationship between the owner and the contractor, the authority of the client's representatives or agents, and the terms of the contract. The *AIA A201* general conditions document is used by many clients and architects.

General conditions (2): *indirect construction costs*, whether in the home office or at the jobsite, that cannot be attributed solely to any direct work activities.

General contractor: the party to a construction contract who agrees to construct the project in accordance with the contract documents; employs direct craft labor and subcontractors; see also *construction manager*.

General contractor/construction manager delivery method: see *construction manager at-risk delivery method*.

Geotechnical report: a report prepared by a geotechnical engineering firm that includes the results of soil borings or test pits and recommends systems and procedures for foundations, roads, and excavation work; also known as a soils report.

Guaranteed maximum price contract: a type of open-book *cost-plus contract* in which the contractor agrees to construct the project at or below a specified cost and potentially share in any cost savings; also guaranteed maximum price estimate.

Heavy civil contractor: the term sometimes used to distinguish general contractors that typically build infrastructures. They usually self-perform more tasks than a commercial GC and often own a fleet of construction equipment.

Histogram: vertical bars rather than a curve reflect the frequency or quantity (such as man-power or cash) of items or occurrences charged against time.

Indirect construction costs or resources: expenses indirectly incurred and not directly related to a specific project or construction activity; see also *general conditions (2)*.

Initial inspection: a quality-control inspection to ensure that workmanship and dimensional requirements are satisfactory; ideally performed before subcontractors demobilize; may also be known as cover inspections.

Integrated project delivery method: fairly new contracting method where the project owner, architect, and general contractor all sign the same contract agreement and share risks equally for financial, safety, schedule, and quality performance.

Internal risk: issues within a company or within their control that they must consider when making plans or taking actions.

Interruptions: scheduling term relating to the stopping and restarting of a construction activity that often leads to inefficiencies.

Invitation to bid: a portion of the bidding documents soliciting bids for a project; also known as instructions to bidders or *request for quotation*.

Jobsite general conditions costs: field indirect costs that cannot be tied to an item of work, but which are project specific, and in the case of cost reimbursable contracts are considered part of the cost of the work; see also *general conditions (2)*.

Jobsite layout plan: two-dimensional plan of the jobsite often authored by the project superintendent that includes locations for hoisting and trailers and site safety and storm water control, among others; see also *site logistics plan*.

Jobsite overhead: see *jobsite general conditions cost*.

Joint occupancy: see *dual occupancy*.

Joint venture: a contractual collaboration of two or more parties to undertake a project or establish a business.

Journeyman: a skilled craftsperson who has advanced through his or her apprenticeship and are qualified to work alone or with the help of an apprentice; they may eventually move on to a master craftsman rank or foreman; see also *craftsmen* and *tradesmen.*

Just-in-time delivery of materials: a material management approach in which supplies are delivered to the jobsite just in time to support construction activities. This minimizes the amount of space needed for on-site storage of materials; an element of *lean construction.*

Labor benefits: voluntary costs paid by contractors on top of employee wages, e.g. union dues, vacation, medical insurance, training, retirement, and others.

Labor burden: addition of *labor benefits* plus *labor taxes.*

Labor taxes: mandatory costs paid by contractors on top of employee wages including FICA, worker's compensation insurance, state and federal unemployment insurance, and Medicare.

Lag: term that connects the start of two activities such that one activity will begin a stated period of time after the first activity; essentially two *start to start* activities with one delayed after the first.

Last planner: individual or group of individuals ultimately responsible for getting the work accomplished, both on the design side and construction side, which is usually a craft *foreman* or *subcontractor*; an element of *lean construction.*

Late finish: the latest an individual activity can finish without delaying the entire project.

Late start: the latest an individual activity can start without delaying the entire project.

Laydown areas: areas of the construction site that have been designated for storage of construction materials that are being stored until they are used in the construction of the project.

Lean construction: a process to improve costs and eliminate waste incorporating efficient methods during both design and construction, includes *just-in-time deliveries, target value design, value engineering, pull-planning*, and other elements.

LEED: Leadership in Energy and Environmental Design; a measure of sustainability administered by the United States Green Building Council, usually associated with receipt of a LEED certificate or plaque.

Letter of intent: a letter, in lieu of a contract, notifying the contractor that the client intends to enter into a contract pending resolution of some restraining factors, such as permits or financing or design completion; sometimes allows limited construction or procurement activities to occur.

Linear schedule method: a graph that measures quantity of work or areas of work or scopes to time, popular with heavy civil construction.

Line item: an entry in the estimate, schedule, or bookkeeping and accounting system.

Link: two or more construction activities are connected; often one is a predecessor of the other. A series of activities which are linked, similar to a chain, are considered a schedule path.

Liquidated damages: an amount specified in the contract that is owed by the contractor to the client as compensation for financial damaged incurred as a result of the contractor's failure to complete the project by the date specified in the contract.

Logic: reasonable order of activities within a schedule that avoids circular or reverse dependencies.

Look-ahead schedule: see *short interval schedule.*

Lump sum contract: a contract that provides a specific price for a defined scope of work; also known as fixed-price or stipulated-sum contract; also lump sum estimate.

Material delivery: the date or time or occurrence when direct materials needed for construction are scheduled to arrive.

Material supplier: a vendor who provides materials but no on-site craft labor.

Meeting notes: a written record of meeting attendees, topics addressed, decisions made, open issues, and responsibilities for open issues.

Microsoft Excel: a spreadsheet utilized for many types of construction schedules, such as *short-internal schedules* or tracking logs.

Microsoft Project: popular scheduling software.

Milestone: a significant or important event or date such as the issuance of the building permit, building dried-in, receipt of certificate of occupancy, and several others.

Mini-schedule: a short schedule or specialty schedule focusing on one area of work or one subcontractor; see also *fragnet.*

Mobilization: the process to physically move onto the construction project before construction commences.

Monte Carlo simulation: a method to predict when activities may occur utilizing statistics and mathematical formulas, focusing on the most probable solution.

Multiple prime project delivery method: when a project owner has qualified in-house staff to manage construction and chooses to contract directly with several specialty contractors; may also be known as five prime delivery method.

Negative float: opposite of float; occurs when the late start and late finish dates are earlier than the early start and early finish dates; discovered during the backwards pass; indicates the project will finish behind the targeted completion date.

Negotiated procurement process: selection of a contractor based on a set of criteria the owner selects; see also *procurement methods* .

Network diagrams: a schedule that shows the relationships among the project activities with a series of nodes and connecting lines, also networks.

Node (1):   as related to arrow-diagramming method: a place or time where two activities meet, usually shown with a circle or square.

Node (2):   as related to precedence-diagramming method: a circle or square that reflects the activity itself and includes information associated with duration, start and finish dates, float, resources, and others.

Notice to proceed: written communication issued by the owner to the contractor, authorizing the contractor to proceed with the project and establishing the date for project commencement.

Occupancy: a point in time when a project owner can utilize their project, often a construction milestone.

Officer-in-charge: general contractor's principal individual who supervises the project manager and potentially the superintendent and is responsible for overall contract compliance.

Off-site prefabrication: prefabrication or assembly of building modules or systems improving on-site cost and schedule performance; also known as off-site construction.

Overlapping schedule: construction activities that do not happen in a pure end-to-end or linear fashion or in series but rather one activity starts before the preceding activity is 100% complete; see also *fast-track schedule*.

Overtime: the amount of time a worker, usually a craftsman, works beyond eight hours in a day in a five-day workweek (unless scheduled to work four 10-hour days), or beyond five days a week; see also *premium time*.

Owner-architect-contractor: three primary parties in any construction agreement who often meet once weekly for an owner-architect-contractor meeting in which schedule and other important topics are discussed.

Pareto Principle: see *eighty-twenty rule*.

Partnering: a cooperative approach to project management that recognizes the importance of all members of the project team; establishes harmonious working relationships among team members, and resolves issues in a timely manner.

Path: a track of continuous events or activities related to one-another; also known as a chain of activities.

Payment request: a document or package of documents requesting progress payments for work performed during the period covered by the request, usually monthly; also known as a pay estimate or invoice.

PERT chart: a scheduling method known as *program evaluation review technique* that utilizes statistics to predict an activity's or project's worst, best, and most-likely durations and dates.

Plan: the up-front work of selecting construction activities and arranging them in a logical order that precedes preparation of the actual schedule document; see also *subcontracting plan, traffic plan, quality control, safety plan,* and *contract schedule*.

Postproject analysis: reviewing all aspects of the completed project, including cost, schedule, quality, and safety performance, to determine lessons that can be applied to future projects.

PowerProject: scheduling software.

Prebid conference or meeting: a meeting of bidding contractors with the project owner and architect. The purpose of the meeting is to explain the project and bid process and solicit questions regarding the design or contract requirements. Usually all contractors are gathered together at one time so that they all hear the same information from the project owner and design team.

Precedence-diagramming method: a technique that uses *nodes* to depict activities and arrows to depict relationships among the activities; used by most scheduling software; also known as activity on the node network or diagram.

Preconstruction agreement: a short contract that describes the contractor's responsibility and compensation for *preconstruction services.*

Preconstruction conference: a meeting conducted by an owner or designer to introduce project participants and to discuss project issues and management procedures.

Preconstruction contract: an agreement indicating services to be provided by the GC, duration of services, and the amount of fee the project owner will pay; also known as a *preconstruction agreement.*

Preconstruction fee: the amount a GC will charge a project owner for costs incurred during the preconstruction period as described in the preconstruction agreement. These costs are typically based on loaded wages (wages plus markups for burden and materials and home office overhead) and often run short of actual costs incurred.

Preconstruction meeting: a weekly or twice-monthly meeting between the design team, project owner, and preconstruction contractor or consultant to discuss design status, schedule, budget options, and other preconstruction services.

Preconstruction phase: time before construction occurs that includes design, permitting, budgeting, scheduling, and other important preconstruction planning activities.

Preconstruction services: services or activities that a construction contractor performs for a project client during design development and before construction starts; includes estimating, scheduling, and constructability reviews.

Predecessor: an activity that precedes or occurs either in part or whole before another activity can commence.

Pre-final inspection: an inspection conducted when the project is near completion to identify all work that needs to be completed or corrected before the project can be considered complete; also known as *punch list.*

Premium time: that portion of a worker's wage that represents the cost of overtime; usually 50% over base wage or double base wage if time is worked on Sundays or holidays, subject to union-contractor negotiations; see also *overtime.*

Preparatory inspection: a quality-control inspection to ensure that all preliminary work has been completed on a project site before starting the next phase of work.

Preproposal conference: a meeting of potential contractors with the project client and architect. The purpose of the meeting is to explain the project, the negotiating process and selection criteria, and solicit questions regarding the design or contract requirements; similar to a *prebid conference.*

Prequalification of contractors: investigating and evaluating prospective contractors based on selected criteria prior to inviting them to submit bids or proposals.

Present value: today's worth of an investment made in prior years or anticipated to be made in future years and adjusted for inflation or interest; see also *future value* and *time value of money.*

Primavera Project Planner or P6: scheduling software system from Oracle, popular on larger construction projects and public work projects.

Prime agreement or contract: a contract entered directly between the project owner (or client or developer) and the general contractor or construction manager.

Procurement schedule: typically a spreadsheet outlining the order in which suppliers and subcontractors will be contracted and material and equipment delivery dates; may also be known as a buyout schedule.

Procurement methods: a project owner will choose to employ a GC either by soliciting competitively priced bids or negotiated proposals and GCs will do likewise with subcontractors and suppliers.

Productivity: a measure of time and efficiency typically associated with direct labor on the jobsite.

Program Evaluation Review Technique: a scheduling method in which calculations and probability are utilized to determine the most optimistic time for a schedule activity to commence, duration, and completion; also *PERT*.

Progress payment request: see *payment request*.

Progress payments: periodic (usually monthly) payments made during the course of a construction project to cover the value of work satisfactorily completed during the previous period.

Project close-out: see *close-out*.

Project control: methods the project team utilizes to anticipate, monitor, and adjust to risks and trends in controlling cost, schedule, quality, and safety.

Project engineer: a project management team member who assists the project manager on larger projects. The project engineer is usually more experienced and has more responsibilities than the field engineer, but less than the project manager. The project engineer is responsible for management of technical issues on the job site; see also *field engineer*.

Project executive: the general contractor's principal individual who supervises the project manager and is responsible for overall contract compliance; see also *officer-in-charge* or *chief operating officer*.

Project item list: a document developed by the GC's estimating team as part of its *work breakdown* effort that lists every work scope and divides it between self-performed or subcontracted; also part of the *subcontracting plan*.

Project labor curve: a plot of estimated labor hours or crew size required per month for the duration of the project.

Project management: the application of knowledge, skills, tools, and techniques to the many activities necessary to complete a project successfully.

Project management close-out: the end-of-project completion of paperwork and documentation including as-built drawings, O&M manuals, test reports, prevailing wage rate schedule, and many others as prescribed in the prime agreement; see also *close-out*.

Project management organization: the contractor's project management group headed by the officer-in-charge, including field supervision and staff.

Project manager: the leader of the contractor's project team who is responsible for ensuring that all contract requirements are achieved safely and within the desired budget and time frame; usually supervises field office staff, including project engineers and jobsite schedulers.

Project manual: a specification volume that also may contain contract documents such as instructions to bidders, bid form, general conditions, special conditions, geotechnical report, prevailing wage rates, and others.

Project owner: the party that the GC typically contracts with and determines the use of the property, establishes entitlement with the AHJ, and secures construction financing; also known as client, customer, developer, or owner.

Project planning: the process of selecting the construction methods and the sequence of work to be used on a project.

Project-specific quality control plan: see *active quality control*.

Project-specific safety plan: a detailed accident prevention plan that is focused directly on the hazards that will exist on a specific project and on measures that can be taken to reduce the likelihood of accidents.

Project start-up: see *start-up*.

Project team: individuals from one or several organizations who work together as a cohesive team to construct a project.

Public-private partnership: a construction delivery method in which a public agency partners with a contractor or developer to reduce costs and lawsuits and ultimately save taxpayer money.

Pull planning: a scheduling method often utilizing sticky notes in which milestones of each design or construction discipline are established and the project is scheduled backwards with the aid of short-term detailed schedules; a tool of *lean construction*.

Pull schedule: lean construction tools prepared by the last planners that have been adapted from the automobile industry; a type of *short-interval schedule*.

Punch list: a list of items that need to be corrected or completed before the project can be considered complete.

Purchase orders: written contracts for the purchase of materials and equipment from suppliers; see also long-form purchase order and short-form purchase order.

Quality control: a process to assure materials and installation meets or exceeds the requirements of the contract documents, also quality control plan.

Quantity take-off sheet: a document or form used by estimators to record counted or measured quantities taken from the drawings during the QTO process and summarized and extended to purchasable units before being transferred to the pricing recap sheets.

Recovery schedule: a schedule revision that changes activity relationships and/or durations in order to meet a fixed project end date.

Relationship: the way two activities interact, such as one preceding or succeeding or restraining another.

Request for information: a document used to clarify discrepancies between differing contract documents and between assumed and actual field conditions; also known as field question.

Request for information log: a spreadsheet for tracking RFIs from initiation through designer response.

Request for proposal: a document containing instructions to prospective contractors regarding documentation required and the process to be used in selecting the contractor for a project.

Request for qualification: a request for prospective contractors or subcontractors to submit a specific set of documents or responses to demonstrate the firm's qualifications for a specific project.

Request for quotation: a request for prospective contractors to submit a quotation for a defined scope of work; also known as instruction to bidders and/or information for bidders.

Resources: necessary or needed supply or support of materials, man-power, cash, or equipment.

Restraint: restrict, control, impedes; typically one activity, milestone, or delivery must be accomplished before another may commence or finish.

Retention: a portion withheld from progress payments for contractors and subcontractors to create an account for finishing the work of any parties not able to or unwilling to do so; often approximately 5% or 10% of the total cost of the work; also known as retainage.

Revenue: total money received by contractors before cost deductions are made.

Risk analysis: identification and acknowledgment of potential downsides of a decision or endeavor, such as an investment or acceptance of a construction contract, and the development of methods to mitigate those risks.

Risk management: a method used to understand project risks and either accept them, mitigate them, transfer them to other parties, or insure against them.

Rolled-up activity: activities in a schedule network may be hidden and reflected by a summary activity; also rolling schedule.

Rough order-of-magnitude cost estimate: a conceptual cost estimate usually based on the size of the project. It is prepared early in the estimating process to establish a preliminary budget and decide whether or not to pursue the project.

Safety plan: a process or document created by a contractor to actively plan to mitigate potential safety accidents.

Schedule: a path of activities or logic diagram presented with a timescale; activities have durations and dates; may be printed on paper or shown on a computer screen.

Schedule control: the act of analyzing the schedule to determine if any area needs to be corrected and subsequently recalculating the schedule to deal with those corrections.

Schedule exception report: a list or summary of activities that have started, are in process, have completed, are ahead of schedule, or are behind schedule; potentially forecasts *negative float*.

Schedule of submittals: a listing of all submittals required by the contract specifications.

Schedule of values: an allocation of the entire project cost to each of the various work packages or systems required to complete a project. Used to develop a cash flow curve for an owner and to support requests for progress payments; also serves as the basis for AIA document G703 or ConsensusDocs 239, which are used to justify pay requests.

Scheduler: the individual on the project team responsible to produce the time-line schedule, may also be the planner, *project manager*, or *superintendent*; see also *lead scheduler*.

Schedule revision: a new issue of a schedule which incorporates status and updates and potentially projects a new completion date.

Schedules (drawings): spreadsheets that utilize rows and columns to provide detail and definition in a more usable format than inclusion on typical drawings; often a Microsoft Excel spreadsheet. These drawing views are popular with all design disciplines.

Schedule status: documenting what activities have started, are in process, or have completed.

Schedule update: a schedule revision to reflect the actual time spent on each activity to date and/or incorporate minor changed conditions, not necessarily a major schedule revision.

Schematic cost or budget estimate: a budget estimate that is prepared at the completion of schematic design.

Schematic design documents: the plans and specifications prepared early in the design process. They typically consist of sketches and preliminary drawings.

Self-performed work: field construction work performed by the general contractor's work force rather than by a subcontractor; also known as direct work.

Semi-detailed cost estimate: an estimate that is prepared during design development that includes estimates for some components based on quantity take-off and estimates for other components based on order-of-magnitude estimates. May also be known as a *guaranteed maximum price*.

Short-interval schedule: a schedule that lists the activities to be completed during a short interval (two to four weeks). Also known as look-ahead schedule or three-week schedule; used by the superintendent and foremen to manage the work.

Short list: the list of best-qualified contractors developed after reviewing the qualification of prospective contractors. Only the contractors on the short list are invited to bid or submit a proposal.

Site logistics plan: a preproject planning tool created often by the general contractor's superintendent that incorporates several elements including temporary storm water control, hoisting locations, parking, trailer locations, fences, traffic plans, etc.; see also *jobsite layout plan*.

Smartsheet: scheduling software.

Special conditions: a part of the construction contract that supplements and may also modify, add to, or delete portions of the general conditions; also known as supplemental conditions.

Specialty contractors: construction firms that specialize in specific areas of construction work, such as painting, roofing, or mechanical; see also *subcontractors*.

Specialty schedule: see *mini-schedules* or *fragnets*.

Specifications: see *technical specifications*.

Staff scheduler: typically a home office specialist whose primary focus is production of construction schedules in support of jobsite teams, may also be the lead scheduler.

Start-to-finish relationship: one activity cannot start until the preceding activity has completely finished.

Start-to-start relationship: a term that correlates two activities starting at the same time.

Start-up: the construction phase after preconstruction and immediately before (or concurrent with) actual physical mobilization onto the jobsite where the superintendent and project manager go about procuring materials and subcontracting strategies and generally develop the organizational structure needed to manage the project.

Start-up checklist: a listing of items that should be completed during project start-up, together with the date of completion for each item.

Start-up log: a spreadsheet used to manage start-up activities relating to suppliers and subcontractors.

Stormwater pollution prevention plan: the process to control storm water runoff often permitted, administered, and inspected by the city or state.

Subcontracting plan: a document prepared by general contractor after a *work-breakdown structure* is ready, which identifies which scopes to include in each subcontract, essentially dividing up the pie among each intended subcontract; may also be known as a *project item list*.

Subcontractor preconstruction meeting: a meeting the GC's project manager and/or superintendent conduct with each subcontractor before allowing him or her to start work on a project.

Subcontractors: specialty contractors who contract with and are under the supervision of the general contractor.

Subcontracts: written contracts between the general contractor and specialty contractors who provide jobsite craft labor and usually material for specialized areas of work; also subcontract agreement.

Submittals: shop drawings, product data sheets, and samples submitted by suppliers and subcontractors for verification by the design team that the materials intended to be purchased for installation comply with the design intent.

Submittal schedule or log: spreadsheet forecasting and tracking *submittal* activity and potentially material delivery; see also *expediting schedule*.

Substantial completion: the state of a project when it has been sufficiently completed such that the owner can use it for its intended purpose.

Successor: an activity that follows another activity; both have a relationship along a particular path within the schedule, the first is the predecessor and the second is the successor; one following activity depends upon the preceding activity.

Summary schedule: an abbreviated version of a detailed construction schedule that may include 20 to 30 major schedule activities or milestones.

Superintendent: the individual from the contractor's project team who is the leader on the jobsite and who is responsible for supervision of daily field operations on the project including quality and safety.

Superintendent diary: see *daily job diary*.

Supplemental conditions: see *special conditions*.

Suppliers: companies that provide materials to a construction project for installation by others, either the GC or subcontractors. The supplier may or may not engage in off-site material fabrication.

Supply chain material management: recognition that costs start at the material fabrication facility and true savings can be saved earlier in the process at that location rather than later during construction installation.

Sustainability: a broad term incorporating many green-building design and construction goals and processes including *LEED*; also sustainable construction.

Tabular schedule: a spreadsheet utilized as a schedule, often a *short-interval schedule* or *milestone* schedule; includes activity descriptions and start and completion dates and durations.

Target value design: a lean construction process where the project budget is established before design begins and each element of the construction project is given a portion of the design, like a piece of pie, and each design discipline must stay within that budget.

Task: see *activity*.

Technical specifications: a part of the construction contract that provides the qualitative requirements for a project in terms of materials, equipment, and workmanship; see also *construction specifications*.

Temporary certificate of occupancy: permission by the city for a project owner to occupy a building; signifies life-safety and building codes are complete but there is still some remaining work, such as landscaping or tenant improvements.

Three-week schedule: also known as look-ahead schedule; see *short-interval schedule*.

Time value of money: an economic study comparing past, present, and future values of money using a series of formulas and spreadsheets and relying on several factors such as interest rates, inflation rates, investment years, and other variables.

Total float: the maximum amount of time one activity may be delayed without delaying the completion of the entire project.

To-do list: a simple list of activities that will happen in the short term.

Top-down scheduling: a method where the project owner provides the GC, or the GC provides a subcontractor a hard completion date without agreeing to a path to achieve that date, different from *collaborative scheduling*.

Touchplan: scheduling software focusing on electronic *pull planning* processes and others.

Tower crane schedule: often a spreadsheet that lists availability for use of the GC-controlled tower crane.

Tradesmen: individuals on the jobsite who accomplish direct construction work; also known as *craftsmen*.

Traditional project delivery method: a delivery method in which the client has a contract with an architect to prepare a design for a project; when the design is completed, the client hires a contractor to construct the project.

Traffic plan: a physical drawing that may include a narrative describing how trucks and deliveries arrive and leave from a jobsite. This is often a building permit requirement.

Unit price contract: a contract that contains an estimated quantity for each element of work and a unit price. The actual cost is determined once the work is completed and the total quantity of work measured; also unit price estimate.

Value engineering: a study of the relative values of various materials and construction techniques to identify the least costly alternatives without sacrificing quality or performance, and/or improving life-cycle costs.

Value engineering log: a spreadsheet that tracks VE proposals developed during the preconstruction phase and records whether they have been accepted or rejected by the project owner and design team or whether additional details or actions are required.

Velocity diagrams: a schedule in the form of a graph where quantities are scheduled and tracked with respect to time, similar to a *linear schedule*; popular with heavy civil projects.

Waterfall schedule: a schedule format in which the first activity is typically located in the top left corner of the schedule and the last activity is located in the bottom right corner of the schedule; may also be known as a stepped or cascading schedule.

Work breakdown structure: a list of significant work items that will have associated cost or schedule implications.

Work days: days when construction can occur; typically does not include weekends, holidays, and days of adverse weather conditions.

Working foreman or working superintendent: a foreman or superintendent, generally on smaller jobsites with less work to supervise, who also works as a craftsperson.

Work package: a defined segment of the work or system required to complete a project.

References

Many of the following resources were referenced in the book and/or utilized for research in its development. Others connect to many facets of construction management, including scheduling. The construction scheduler or construction management enthusiast may want to refer to some of these other resources.

AGC's Supervisory Training Program, *Unit 3: Planning and Scheduling* (2015 Edition). Arlington: AGC.

AIA Document A101 – 2017 Standard Form of Agreement Between Owner and Contractor where the basis of payment is a Stipulated Sum.

AIA Document A102 – 2017 Standard Form of Agreement Between Owner and Contractor where the basis of payment is Cost of the work Plus a Fee with a Guaranteed Maximum Price. Retrieved 2019 from American Institute of Architects [AIA]. http://aiad8.prod.acquia-sites.com/sites/default/files/2017-07/A102_2017.sample2.pdf

AIA Document A201 – 2017 General Conditions of the Contract for Construction. Retrieved 2019 from American Institute of Architects [AIA]. https://www.aiacontracts.org/contract-documents/25131-general-conditions-of-the-contract-for-construction.

ConsensusDocs® 500, Standard Agreement and General Conditions between Owner and Construction Manager (where the CM is at-Risk), 2015. Contract association which replaced the AGC contracts. https://www.consensusdocs.org.

ConsensusDocs® 751, Standard Short From Agreement between Constructor and Subcontractor, 2014.

Glavinich, T. (2017). *Construction Planning and Scheduling* (2nd ed.). The Associated General Contractors of America.

Hinze, J. (2011). *Construction Planning and Scheduling* (4th ed.). Upper Saddle River, NJ: Pearson.

Holm, L. (2019). *101 Case Studies in Construction Management.* New York: Routledge.

Holm, L. (2020). *Construction Contract Documents, Including Plan Reading Essentials and Extensive Lists of Abbreviations and Construction Glossary Terms.* Amazon.

Holm, L., & Schaufelberger, J. (2021). *Construction Cost Estimating.* New York: Routledge.

Holm, L., & Schaufleberger, J. (2020). *Construction Superintendents, Essential Skills for the Next Generation.* New York: Routledge.

Holm, L. (2019). *Cost Accounting and Financial Management for Construction Project Managers.* New York: Routledge.

Kim, Yong-Woo. (2017). *Activity-Based Costing for Construction Companies.* Oxford, UK: Wiley.

Lean Construction Institute (https://www.leanconstruction.org).

Merriam Webster's Collegiate Dictionary (2020, 23rd printing, 11th ed.).

Migliaccio, G., & Holm, L. (2018). *Introduction to Construction Project Engineering.* New York: Routledge.

Mubarak, S. (2019). *Construction Project Scheduling and Control* (4th ed.). Hoboken, NJ: Wiley.

Pierce, D. (2013). *Project Scheduling and Management for Construction* (4th ed.). Hoboken, NJ: Wiley.

Popescu (1995). *Project Planning, Scheduling, and Control in Construction*, Encyclopedia of Terms. Wiley.

Schaufelberger, J., & Holm, L. (2017). *Management of Construction Projects, A Constructor's Perspective* (2nd ed.). New York: Routledge.

"The Plan for the Settlement of Jurisdictional Disputes," revised May 1, 2011, https://nabtu.org/field-service/plan-settlement-jurisdictional-disputes), also known as The Green Book.

Weber, S. (2005). *Scheduling Construction Projects: Principles and Practices. Upper Saddle River*, NJ: Pearson.

Five Sample Case Studies

Following are a few select cases chosen from *101 Case Studies in Construction Management* related to planning, scheduling, and controls. The numbering of the cases is the same numbering as that utilized in the published case study book. This information is copyrighted; please respect the author's work and not copy it. The case study book is an excellent and inexpensive accompaniment for any course in construction management or construction engineering. A separate instructor's guide with possible solutions, titled *Suspects*, is available gratis to university instructors. The instructor's guide also suggests several possible uses for the cases in the classroom for the benefit of the instructor. Industry professionals may contact Len Holm at holmcon@aol.com for an instructor's guide if interested.

CASE 40: GLAZING SCHEDULE

You, as the general contractor's (GC's) project manager (PM), have a problem with a glazing subcontractor. They are behind schedule. They refuse to work overtime to catch up. The subcontractor has submitted several unsubstantiated change order proposals (COPs) that have not yet been approved and they are threatening to stop work. They have switched out both the project manager and the superintendent since the project started. They are not staffing the project according to their planned and committed manpower. You are not getting along personally with the subcontractor's current project manager and have resorted to communicating only through email. You are receiving pressure from the field to resolve the problem and get the glazier to perform. Your supervisor has indicated that it is your responsibility to solve the problem. What do you do? What could you have done to prevent these problems from occurring? What recommendations can you make to a general contractor's subcontractor management system to prevent these types of situations?

_____

_____

_____

_____

_____

_____

_____

CASE 41: DRYWALL SUBCONTRACTOR

Your drywall subcontractor is not performing in the field. They have not staffed the project according to their original commitment or to your field superintendent's expectations. They are holding up the work of other trades. The quality of the subcontractor's completed work has been unacceptable and you are constantly on them to improve. You have asked for the removal of their superintendent but the firm has refused. Can you contractually require a subcontractor to change personnel? Can they contractually refuse? It is eventually decided by your home office that the subcontractor must be terminated. How do you go about this process? Is it simple? How is it documented? Will your firm get sued for false termination? What does standard contract language say? How do you protect yourself? Will it be easier to just keep limping along with them? Would your answers differ if your GC firm had either a lump sum or a negotiated contract with your client?

_____

_____

_____

_____

_____

_____

CASE 42: LIQUIDATED DAMAGES

Your client has assigned $2,000 per day liquidated damages for late completion to your $25 million turn-key contract. You in turn have passed these liabilities on to your subcontractors. You have one exterior siding subcontractor who sends you written notices requesting additional time whenever a request for information (RFI) or a submittal is a day late being returned. They document every adverse weather day. They document when other subcontractors are holding them up. They request schedule extensions with every change order. Not all of their documentation is substantiated, but some of it may be. They are claiming a total of 20 additional work days. This is an administrative nightmare for you.

a. How do you deal with the siding contractor during the course of the project? Do you pass through these notifications to your client? If this subcontractor ultimately finishes behind schedule by 10 days, and you also finish behind schedule, does your client collect from you? Do you collect from the subcontractor? How does it get resolved typically? What is the contractual and legal resolution? Analyze both ways: (A) with stipulated LDs in the subcontract agreement, and (B) without.

_____

_____

_____

_____

_____

_____

_____

b. Do you offer a subcontractor a bonus ($/day) for finishing early? Isn't this only fair? If the penalty were $2,000 per day for finishing late, what would the bonus be? Assume the siding contractor above finished per their original schedule, but had built up the claimed additional 20 days, and had such a bonus clause. They would now claim that a bonus was due for the $40,000. Do they get it? If not, why not? If so, who pays; you or your client? How is this resolved?

_____

_____

_____

_____

_____

_____

_____

CASE 43: SCHEDULE HOLD

Your construction company does $200 million in annual volume with a home office overhead cost amounting to a total of 2% of annual revenue. This particular 16-month project has costs estimated at $20 million plus 8% jobsite general conditions costs plus a 4% fee. Ignore other markups for this exercise. Assume that at exactly the mid-point of the schedule your project was put on hold for one month due to reasons beyond the contractor's control such as weather, union strikes, owner financing, or city issues. Pick one.

a. Using the contract, classes, your book, and research outside of the classroom, how should the contractor properly deal with this delay? Discuss issues such as notice, documentation, jobsite administration costs, home office costs, loss of fee, loss of productivity, and quality and safety concerns. What is the "Eichleay Formula"? Prepare a claim for this delay.

_____

_____

_____

_____

_____

_____

_____

b. As an alternative to the claim preparation above, prepare a recovery plan to get the owner their building on time. Show with the schedule that a recovery is possible. Submit a change order proposal for the anticipated costs associated with working in an expedited fashion. Provide all necessary cost backup.

_____

_____

_____

_____

_____

_____

CASE 77: SUBCONTRACTOR QUALITY CONTROL

This general contractor was constructing a very sensitive medical tenant improvement (TI) project. The superintendent was very laid back and the project manager (PM) was at times combative with the client and the architect. The architect noted two quality control concerns early in the project that warranted more attention. This was the preparation and flashing of roof penetrations and the taping and fireproofing of the vertical drywall surfaces. These conditions occurred around a medical procedure room that required close attention of all the team members. The superintendent's response to the drywall issues was "the subcontractor is the expert and I don't tell him what to do." The PM further responded to the roof issues with "unless you want to pay me a change order to change the process, we will stay with our course. If it leaks, the roofing subcontractor will have to return and incur the cost of repairs." Is this a client-friendly GC? Are they contractually correct? The architect is trying to implement active QC by bringing potential problems up early. Is this a good practice? Given the GC's response, will the architect continue with early notifications? If this is a "means and methods" issue, can the architect force changes without a change order? Can she issue an AIA CCD to force the issue? If the architect's concerns go uncorrected but eventually turn out to be a warrantee issue, is the owner in a better position to claim impact against the general? Will the owner or architect choose this GC again?

_____

_____

_____

_____

_____

_____

_____

Index